Artificial Tactile Sensing in Biomedical Engineering

About the Authors

Prof. Siamak Najarian has completed his Ph.D. in Biomedical Engineering at Oxford University, England and had a pos-doc position at the same university for one year. Prof. Najarian serves as the Full-Professor and Dean of Faculty of Biomedical Engineering at Amirkabir University of Technology. His research interests are the applications of artificial tactile sensing (especially in robotic surgery) and design of artificial organs. He is the author and translators of 23 books in the field of biomedical engineering, 8 of which are written in English. Prof. Najarian has published more than 130 international journal and conference papers in the field of biomedical engineering.

Prof. Javad Dargahi received his B.Sc., M.Sc., and Ph.D. degrees in mechanical engineering in the UK. He worked as a research assistant and completed his Ph.D. degree in robotic tactile sensing at Glasgow Caledonian University. He was a senior postdoctoral research associate with the micromachining/medical robotics group at Simon Fraser University. Dr. Dargahi was employed as an assistant professor in the biomedical engineering department at Amirkabir University of Technology and also as a full-time lecturer at University of New Brunswick. He worked for the Pega Medical Company in Montreal before he joined Concordia University, where he is an associate professor. His research interests are design and fabrication of haptic sensors and feedback systems for minimally invasive surgery and robotics. He has published over 135 journal and conference papers and owns a teletaction patent. Dr. Dargahi was also a reviewer of several NASA and NSERC proposals.

Ali Abouei Mehrizi holds a master's degree in biomedical engineering from Amirkabir University. He coauthored a book in the field of biomedical engineering and has written articles on the subject for accredited international journals and conferences.

Artificial Tactile Sensing in Biomedical Engineering

Siamak Najarian
Javad Dargahi
Ali Abouei Mehrizi

New York   Chicago   San Francisco
Lisbon   London   Madrid   Mexico City
Milan   New Delhi   San Juan
Seoul   Singapore   Sydney   Toronto

The McGraw·Hill Companies

Library of Congress Cataloging-in-Publication Data

Najarian, Siamak.
  Artificial tactile sensing in biomedical engineering / Siamak Najarian, Javad Dargahi, Ali Abouei Mehrizi.
     p. ; cm.
  Includes bibliographical references and index.
  ISBN 978-0-07-160151-1 (alk. paper)
  1. Tactile sensors—Congresses. 2. Robotics in medicine—Congresses.
  I. Dargahi, Javad. II. Mehrizi, Ali Abouei. III. Title.
  [DNLM: 1. Diagnostic Techniques and Procedures—instrumentation.
  2. Touch Perception. 3. Biomedical Engineering. 4. Robotics. WL 702
  N162a 2009]
  R857.T32N25 2009
  610.28'4—dc22

                                                                  2009010122

Copyright © 2009 by The McGraw-Hill Companies, Inc. All rights reserved. Printed in the United States of America. Except as permitted under the United States Copyright Act of 1976, no part of this publication may be reproduced or distributed in any form or by any means, or stored in a data base or retrieval system, without the prior written permission of the publisher.

1 2 3 4 5 6 7 8 9 0   DOC/DOC   0 1 4 3 2 1 0 9

ISBN: 978-0-07-160151-1
MHID: 0-07-160151-1

The pages within this book were printed on acid-free paper.

Sponsoring Editor Taisuke Soda	**Copy Editor** Romi Sussman	**Composition** Value Chain International Ltd
Acquisitions Coordinator Michael Mulcahy	**Proofreader** Julie Grady	**Art Director, Cover** Jeff Weeks
Editorial Supervisor David E. Fogarty	**Indexer** Julie Grady	
Project Manager Julie Waldman	**Production Supervisor** Pamela A. Pelton	

Information contained in this work has been obtained by The McGraw-Hill Companies, Inc. ("McGraw-Hill") from sources believed to be reliable. However, neither McGraw-Hill nor its authors guarantee the accuracy or completeness of any information published herein, and neither McGraw-Hill nor its authors shall be responsible for any errors, omissions, or damages arising out of use of this information. This work is published with the understanding that McGraw-Hill and its authors are supplying information but are not attempting to render engineering or other professional services. If such services are required, the assistance of an appropriate professional should be sought.

Contents

Preface		xi

**1 The Four Senses in Humans:
Sight, Hearing, Smell, and Taste** **1**
   1.1 Sense of Sight 1
           The Adjustment in the Eyes 2
   1.2 Sense of Hearing 6
           Sound Waves 8
           The Function of the Outer Ear 8
           The Function of the Middle Ear 9
           The Function of the Inner Ear 9
           The Function of the Round Window 10
           The Activation of Auditory Afferents 10
           The Pattern of Vibration of
               the Basilar Membrane 11
           The Coding of Frequency of a Sound 11
           The Coding of Loudness 11
           Hearing Loss 12
   1.3 Sense of Smell 12
           The Olfactory System 13
   1.4 Sense of Taste 16
           References 18

2 The Sense of Touch **19**
   2.1 Introduction 19
   2.2 The Exteroceptive Sensor System 22
           The Receptive Field 24
   2.3 The Proprioceptive Sensor System 24
   2.4 Transduction of Mechanical
           Stimuli to Neural Impulses 27
   2.5 Pathways of Tactile Information 30
   2.6 Special Features of Tactile Sensing 31
           References 33

**3 Introduction to Tactile Sensing and
Tactile Sensors** **35**
   3.1 Tactile Sensing 35
   3.2 Tactile Sensors 38
           Terminology of Artificial Sensors 38
           Resolution 39

vi Contents

		Transfer Function	39
		Sensitivity	39
		Calibration	40
		Linearity	40
		Hysteresis	40
		Accuracy	41
		Span or Dynamic Range	42
		Noise	42
		Repeatability	43
		Reliability	43
		Response Time	43
		Some Other Specifications for Tactile Sensors	44
		Classification of Tactile Sensors	44
		References	47
4	**Introduction to Tactile Sensing Technologies**		**49**
	4.1	Introduction	49
	4.2	Capacitive Sensors	49
	4.3	Inductive Sensors	52
		Linear Variable Differential Transformer (LVDT)	55
	4.4	Conductive Elastomers and Carbon Fibers	59
	4.5	Optical Sensors	63
	4.6	Thermal Sensors	65
	4.7	Time of Flight Sensors	65
	4.8	Binary Pressure Sensors	66
	4.9	Fluidic Coupling	68
	4.10	The Hall Effect and Magnetoresistance	68
		References	71
5.	**Strain Gauge Sensors**		**73**
	5.1	Introduction	73
	5.2	Metal Strain Gauges	73
	5.3	Semiconductor Strain Gauges	81
		References	84
6	**Piezoelectric Sensors**		**85**
	6.1	Piezoelectric Materials	85
	6.2	Piezoelectric Ceramics	85
	6.3	Directional Dependence of Piezoelectricity	86
	6.4	Polyvinylidene Fluoride	91

6.5 Piezoelectric Sensors in
    Biomedical Applications 91
    A Piezoelectric Tactile Sensor
      for Use in Endoscopic Surgery 92
    A Multifunctional PVDF-Based
      Tactile Sensor for Minimally
      Invasive Surgery 92
    A Piezoelectric Tactile Sensory
      System with Graphical
      Display of Tactile Sensing Data 94
    A Hybrid Piezoelectric-Capacitive
      Tactile Sensor 99
    References 103

7 Application of Tactile Sensing in Surgery 105
7.1 Open Surgery and Minimally
    Invasive Surgery 105
7.2 Basic Components of a Tactile
    Sensing System for Use in MIS 108
    Tactile Sensor 108
    Tactile Data Processing 109
    Tactile Display 111
    Design Considerations for
      Tactile Sensing Systems in MIS 111
7.3 Remote Palpation Instruments for MIS 112
    Design Specifications for
      Remote Palpation Instruments 114
    Analysis of Contact Force
      Between an Endoscopic
      Grasper Used in MIS and the
      Biological Tissues 115
    References 122

8 Tactile Image Information 123
8.1 Introduction to Palpation 123
8.2 Taxonomy of Palpation 124
8.3 Palpation and Tactile Image 125
    Information for Mapping
      Tactile Imaging 125
    Imaging Procedures for
      Breast Cancer 130
    Breast Self Exam 130
    Clinical Breast Exam 131
    Mammography 131

	Tactile Imaging and Breast Cancer Screening	131
	Estimating of Lesion Parameters	132
	Analytical Solution	133
	Tactile Information from Finite Element Models	135
	Inversion Algorithm	136
	References	140
9	**Application and Recent Developments of Tactile Sensing in Tumor Detection**	**143**
9.1	Introduction	143
9.2	Detection of Tumors Using a Computational Tactile Sensing Method	143
9.3	Application of Artificial Neural Networks for the Estimation of Tumor Characteristics in Biological Tissues	149
9.4	Prediction of Tumor Existence in the Virtual Soft Tissue by Using Tactile Tumor Detector	152
9.5	Graphical Rendering of Localized Lumps for MIS Applications	153
	System Design	155
	Sensor Structure	155
	Rendering Algorithm	156
	Experiments	164
	Results	166
	References	169
10	**Determination of Mechanical Properties of Biological Tissues Including Stiffness and Hardness**	**171**
10.1	Introduction	171
	Determining the Stiffness of Cartilage	172
	Tactile Sensor System	172
10.2	Experimental and Theoretical Analysis of a Novel Flexible Membrane Tactile Sensor	173
	Sensed Objects	173
	Two-Dimensional Surface Texture Image Detection	174
	Contact-Force Estimation	174
	Stiffness Detection	174

		Device Specification	175
		Theoretical Analysis	175
		Experimental Method	176
		Results	178
	10.3	A Micromachined Active Tactile Sensor for Hardness Detection	180
		Principle of the Tactile Sensor	180
	10.4	Design and Fabrication of a New Tactile Probe for Measuring the Modulus of Elasticity of Soft Tissues	182
		Introduction	183
		Description of the System	183
	10.5	Tactile Distinction of an Artery and a Tumor in a Soft Tissue by Finite Element Method	184
		Materials and Methods	186
		Results	187
	10.6	Artificial Skin	193
		References	195
11	**Application of Tactile Sensing in Robotic Surgery**		**197**
	11.1	Robot Definitions	197
		An Aspect of an Integrated System	197
	11.2	Application of Robots in Surgery	198
		Robotics in Surgery	199
		Current Applications of Robotic Surgery	200
		Suturing in MIS	201
		Laparoscopic Suturing	202
		Tension Measurement in Suturing	203
		Commercial Robots for Surgery	205
		Companies Which Produce Commercial Robots	205
		Commercial Robots for Surgery	205
	11.3	Robots for MIS	208
		Force Sensors for Surgical Robots	208
		Teleoperation	209
		Telemonitoring Skin Conditions	211
		A Tactile Sensor for Detection of Skin Surface Morphology	212
		References	218

Contents

12 Haptics Application in Surgical Simulation 221
  12.1 Virtual Reality (VR) and Virtual Environments (VEs) 221
    Applications of Virtual Reality 222
    Advantage and Limitation 222
  12.2 Haptics-Based Surgical Simulation 223
    Medical Training Simulation 224
    Deformable Models for Tissue Simulation 225
    Haptic Simulation 226
    Fluid Simulation 227
    Surgical Simulators Based on Haptics 227
    Needle-Based Procedure 228
    References 229

Abbreviations 231

Index 233

Preface

There are five types of sensing modalities, namely touch, sight, smell, sound, and taste. To understand the development of the tactile sensing system, it is useful to compare these types of senses. For the sense of sight, artificial vision systems are now widely used. This is also true for many sound or audio systems. In this category, speech analysis and recognition topics are among the most active research areas. Audio systems have found a number of applications in various industries. Nowadays, even the studies on the senses of smell and taste are moving towards devices that can, respectively, act as an artificial nose or taste simulator. Generally, the five sensing modalities for a healthy human-being are, in fact, completely different devices in structure and physiology. These senses act as information paths entering from the outside world to the human brain for a better understanding of the environment. In contrast to other senses, the sense of touch system is unevenly spread throughout the entire body; indicating the special importance of this sense with respect to other sensing modalities.

Tactile sensation is the process of determining physical properties and events of an object by making contact with that object. A binary system is quite similar to the nature of touch sensing insofar as it deals with either the existence of contact or the lack of it. In this regard, stimuli encountered in tactile sensing systems vary from determining simple status of contact to obtaining a complete map of the state of touch.

The development of tactile sensing began in the 1970s; since that time, it has proven useful and efficient in many different fields such as industrial robot grippers, haptic perception, and multifingered hands for dexterous manipulation. At the time, most researchers predicted that the main application for tactile sensing would be industrial automation activities and, to target this goal, several commercial tactile devices were introduced. However, the demand for such devices (and, hence, the number sold) was less than expected, possibly due to their limitations. Nowadays, however, it is believed that there are many new fields where tactile sensing is likely to play a key role. Among these novel fields are: medical procedures (especially surgery); rehabilitation and service robotics; and agriculture and food processing.

Preface

In the medical field, surgery is the most rapidly developing area where tactile sensing is becoming crucially important. Although Minimally Invasive Surgery (MIS) is a new field in surgical procedures, this type of surgery is expected to become the preferred choice for surgeons around the world. However, despite its advantages, MIS severely reduces the surgeon's sensory perception during manipulation. This is because surgery is essentially a visual and tactile procedure, and any limitations on the surgeon's sensory abilities can be undesirable.

This unique book, entitled *Artificial Tactile Sensing in Biomedical Engineering* is a result of twelve years of teaching and research conducted by its authors and other researchers in different universities and research centers. This book is mainly prepared for graduate students and researchers in the field of tactile sensors, with its focus on surgical and diagnostic applications. It includes twelve chapters with more than two-hundred illustrations, presenting practical and fundamental concepts in the field of tactile sensing. The goals and main concepts provided in each chapter are as follows:

Chapter 1 defines various senses in human beings. It starts by introducing the sense of sight and its components, the different receptors and their performance, and a brief anatomical and physiological description. This is followed by the sense of hearing, detailing the structure and the performance of the inner, middle, and outer ear, together with recognition and segregation of air waves. Next is a description of the sense of smell and its perception, followed by an anatomical description of the nasal cavity. Finally, the sense of taste is investigated, including how it performs and recognizes the four main tastes.

Chapter 2 briefly describes the human skin and its tasks, and then focuses on tactile sensing in its different forms (exteroceptive and proprioceptive sensor systems) with a final mention on its performance and special capabilities with respect to other senses. The allocation of one chapter to the sense of touch follows the aim of this book; elucidating a full recognition of tactile sensing and its applications in different medical procedures, surgical operations, and manufacturing synthetic samples in the form of artificial tactile sensing systems.

Chapter 3 commences by introducing haptic and kinesthetic concepts, together with a description of tactile sensing. This is followed by an introduction on the definition of tactile sensors and their classification. Finally, different design features which should be considered in sensor manufacturing are described.

Chapter 4 explains the principles and properties of capacitive sensors, inductive sensors, conductive elastomer and carbon fiber sensors, optical sensors, thermal sensors, time-of-flight sensors, binary pressure sensors, fluidic coupling and, finally, the Hall effect and magnetoresistance sensors.

Chapters 5 and 6 focus exclusively on two types of sensors, the strain gauge and piezoelectric sensors. This focus is due to the importance of these two sensors in the manufacturing of tactile sensors.

Chapter 7 investigates the application of tactile sensing in surgery and palpation. First, the two main categories of surgery (open surgery and minimally invasive surgery) are discussed. The role of tactile sensing is compared in these two methods, together with the advantages and disadvantages of each. Following this, we describe the components of a tactile sensing system and its design features for a minimally invasive surgical procedure. Then, we address remote palpation instruments and the properties which should be considered in various designs. Finally, analysis of the contact force between an endoscopic grasper used in MIS and the biological tissues is investigated as an example.

Chapter 8 deals with tactile image information. Due to their importance in identifying cancerous tumors, especially in breast cancer, palpation and taxonomy of palpation are described in detail, as are as other methods of cancer diagnosis systems. At the end of this chapter, examples are given of research work on modeling breast tumors, extraction of results for recognizing the presence of a tumor from tactile maps, and estimation of tumor parameters using numerical and analytical methods.

Chapter 9 classifies theoretical and experimental research activities conducted in the field of diagnosis of cancerous breast tumors with the aid of artificial tactile sensing. We then show how the parameters of a tumor can be estimated by using the finite element method and an artificial neural network.

Chapter 10 presents various research studies conducted by the authors for determining mechanical properties of soft tissue, such as elastic modulus and stiffness, with the help of different tactile sensors, such as the membrane tactile sensor. Also, for the first time, the detection of embedded arteries is introduced using the finite element tactile modeling method.

Chapter 11 describes the application of tactile sensing in remote and robotic surgery. This chapter primarily introduces robotic related concepts such as why their use, as opposed to that of a human, is preferred. Also mentioned are commercial robots used in various surgeries and their properties. Additionally, the problems pertaining to suturing in MIS, different kinds of sutures, and suturing methods are fully described. Finally, the capability of tactile sensing in diagnosing cutaneous conditions, as well as several research reports carried out in this field, are presented briefly.

Chapter 12 explains the role and importance of tactile sensing in manufacturing surgical simulators. The concepts of virtual environment, along with the advantages and disadvantages of virtual reality systems, are then investigated.

The authors would like to thank the many people who have contributed to this book. We are indebted to our colleagues who have shared their expertise with us, given permission to share their work, and contributed so much to the progress of biomedical engineering students and various medical professions. We also want to appreciate and acknowledge all personal and organizational help provided by various institutes or companies for allowing us to use their materials free of charge.

SIAMAK NAJARIAN
Professor of Biomedical Engineering
Faculty of Biomedical Engineering
Amirkabir University of Technology

JAVAD DARGAHI
Associate Professor of Mechanical Engineering
Department of Mechanical and Industrial Engineering
Concordia University

ALI ABOUEI MEHRIZI
Research Associate of Biomedical Engineering
Faculty of Biomedical Engineering
Amirkabir University of Technology

CHAPTER 1
The Four Senses in Humans: Sight, Hearing, Smell, and Taste

Immanuel Kant, the well-known philosopher, proposed that our knowledge of the outside world depends on our modes of perception. Before discussing extrasensory, we must understand and describe the term *sensory*. There are five known senses available to human beings: touch, taste, smell, hearing, and sight. They all have specialized cells which, in turn, contain receptors for particular stimuli and from which the primitive process of sensing occurs. They are all connected to the nervous system where all sensations are integrated and then linked to the brain.

Another definition of a sense is given in the Wikipedia website as follows: "It is a system that consists of a group of sensory cell types that responds to a specific physical phenomenon, and that corresponds to a particular group of regions within the brain where the signals are received and interpreted."

In this chapter, we examine and explain all of these senses except the sense of touch. Since the sense of touch is our main focus, more details of this sense are provided separately in Chap. 2.

1.1 Sense of Sight

The sense of sight or vision describes two processes. It involves the ability to detect light waves within the visible range by organs called eyes, and the ability of the brain to interpret the received information as an image or what is called sight. This sense allows us to assimilate information from the environment to guide our decisions and actions.

Eyes play an essential role in the process of acquiring visual information. Figure 1.1 shows the structure of this organ. The act of seeing can be briefly described by the following mechanism:

1. The eye lens focuses the light emitted by objects in the outside world onto a light-sensitive membrane in the back of the eye, called the retina.
2. The retina membrane is a multilayered sensory tissue containing millions of photoreceptor cells, serving as a transducer for the conversion of patterns of light into neuronal signals. These cells detect the photons of light and respond by producing neural impulses.
3. A network of neurons collects these impulses and transmits them down the optic nerve to the primary and secondary visual cortex of the brain. The brain then interprets these impulses into vision.

It is by this mechanism that we can explore a variety of apparent characteristics of surrounding objects. These characteristics include shape, color, quantity, and more. Some of the main characteristics are summarized in Table 1.1.

The Adjustment in the Eyes

As described, the act of seeing starts when the lens focuses rays of light on the retina. The lens is a transparent structure in the eye; its main function is to change the focal distance of the eye, allowing a person to focus on objects at various distances. This adjustment of the lens is also known as accommodation. Adjustment is controlled by the nervous system and executed by a system of donut-shaped muscles around the lens called the ciliary muscles. The average lens is about 5 mm thick and has a diameter of about 9 mm for an adult human. Its thickness can be adjusted by the ciliary muscles; crystallins

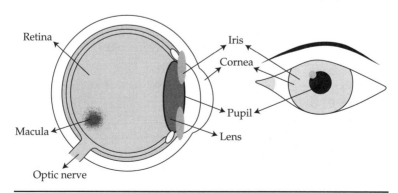

FIGURE 1.1 The structure of the eye.

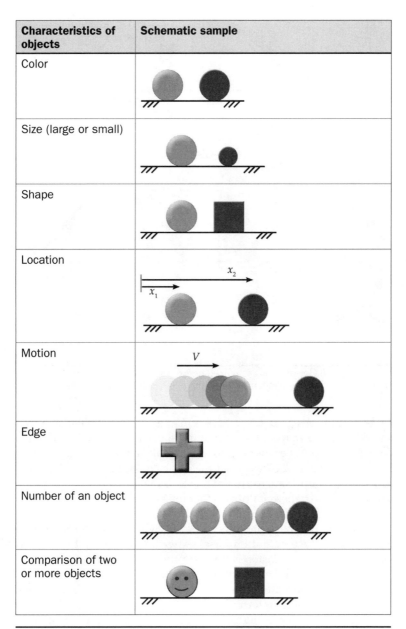

TABLE 1.1 Features Explored by the Eyes.

are its main protein building blocks. Figure 1.2 shows the placement of the lens in the eye structure and the function of its adjustment by the ciliary muscles.

The ability to clearly focus an image depends on two main factors:[1] the shape of the eyeball and the adjustment capabilities

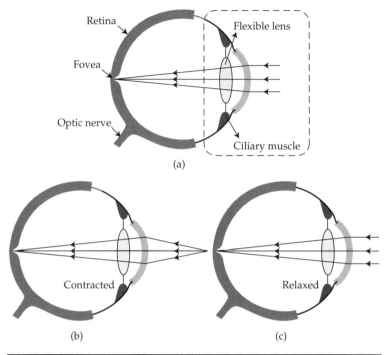

Figure 1.2 (a) How the lens is placed before the cornea in the eye structure, (b) contraction of the ciliary muscles, and (c) relaxation of the ciliary muscles.

of both the lens and the ciliary muscles. If the eyeball is in an out-of-shape form (e.g., too long or too short as in Fig. 1.3a and b) the focal point of the eye lens may not exactly fall on the retina membrane. The same would also happen if the ciliary muscles fail to adjust the lens in an appropriate form, as shown in Fig. 1.3c and d. One of the consequences of these failures is that the person might not be able to clearly see near or far objects.

The retina contains photoreceptor cells whose function is to detect light and produce neural impulses. After a beam of light is focused on the retina, the next step is transducing these patterns of light into neural impulses so that the brain can interpret them. There are two main types of photoreceptor cells in the retina: rod cells and cone cells, which are often referred to simply as rods and cones.[2]

Named for their cylindrical shape, rods are highly sensitive to light; they allow us to see at lower levels of illumination and they are responsible for night vision. Rods are concentrated at the outer edges of the retina and are used in peripheral vision. There are approximately 125 million rod cells in the human retina.[2] A rod cell is sensitive enough to respond to a single photon of light, and is about 100 times more sensitive to a single photon than cones. However, they are unable to fully resolve fine details and are subject to saturation.

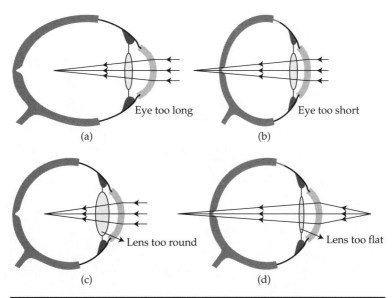

FIGURE 1.3 Some common failures in focusing an image on the retina by the eye: (a) eye too long, (b) eye too short, (c) lens too round, and (d) lens too flat.

This is why we experience a temporary blindness when we move from a dark area into a bright area.

Cones, on the other hand, are less sensitive to light and require tens to hundreds of photons to become activated. There are approximately 6.5 million cones[2] that are mainly concentrated on the fovea, a small area in the center of the retina where images are formed.

There are three types of cones. Each type is sensitive to a different wavelength of light and is, therefore, responsible for a different color vision: red, green, or blue. Other colors are a result of the combined stimulation of these three cones. For example, as Fig. 1.4 shows, when the cones responsible for red and green colors are equally stimulated, a person sees yellow. When all three cones are equally stimulated, a person sees white.

A third category of photosensitive cells also exists in the retina, which is not involved in vision. A small proportion of the ganglion cells, about 2% in humans, responds primarily to blue light, whose wavelength is approximately 470 nm. The signals from these cells do not go through the optic nerve, and can, therefore, function in totally blind individuals.

Now that the impulses are produced by rods and cones, they should be collected and converged. This is mainly done by four types of cells in the eye's neural structure:

1. There are horizontal (H) cells which converge signals from several cones.

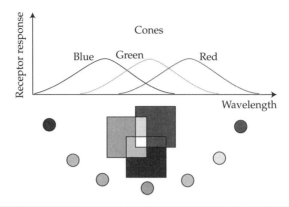

FIGURE 1.4 Colors created by the combination of cones responsible for three main wavelengths (colors). (See also color insert.)

2. There are bipolar (B) cells which connect the receptors to the ganglion cells.
3. There are amacrine (A) cells which converge signals from peripheral rods.
4. Finally, there are ganglion (G) cells which receive, collect, and converge all of the neural signals. They produce action potential and are the only output pathway of neural signals from the eye. The ganglion fires action potentials (AP) at a rate of 10 Hz. This is an AP every 100 ms.

A schematic of cell order and how they are connected to each other is shown in Fig. 1.5.

Finally, the brain interprets these impulses and constructs the image. The images from the two eyes are combined in the primary visual cortex, located in the back of the brain as shown in Fig. 1.6a. The primary visual cortex is a plate of cells 2 mm thick, with a surface area of a few square inches that contains approximately 200 million cells. The visual information from each eye is carried to the primary visual cortex separately. Receiving separate information from both eyes at the same time enables the primary visual cortex to compute the three-dimensional aspect as well. The primary visual cortex separates the image into distinct feature channels and the information is sent to the other visual-related areas of the brain. Different groups of cells work collectively to extract each feature such as color, depth, and edges (Fig. 1.6b).

1.2 Sense of Hearing

Hearing (or audition) is one of the traditional five senses. It is the ability to perceive sound by detecting it via an organ called the ear.

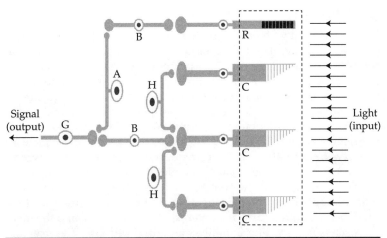

FIGURE 1.5 The cells that produce, collect, and converge the neural impulses and their connections. R is a rod cell and C is a cone cell.

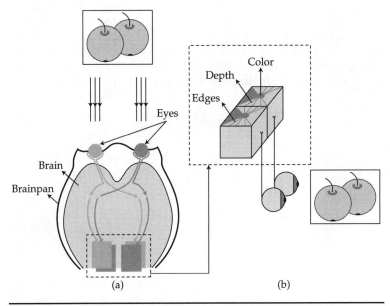

FIGURE 1.6 (a) Images seen by both eyes are interpreted in the left and right cortex. (b) Features such as color, depth, and edges are extracted by a different group of cells. (See also color insert.)

Sound waves are caused by the vibrations of air; these vibrations are detected by the ear and converted into nerve impulses that are perceived by the brain. The auditory system is the sensory system for the sense of hearing.

Sound Waves

Sound waves, by their nature, are waves of air pressure. If one plots air pressure, then the peaks correspond to points of maximum compression, and the troughs correspond to points of maximum rarefaction. Amplitude is the difference between minimum and maximum pressure and is perceived as loudness. Frequency is the number of peaks that go by a fixed point in one second and is perceived as modulation. The normal range of frequencies audible to humans is 20 to 20,000 Hz.[3] We are most sensitive to frequencies between 2000 to 4000 Hz,[1] the frequency range of spoken words. Figure 1.7 illustrates these characteristics of a sound wave.

The ear organ is responsible for detecting such waves and transmitting them as neural signals to the brain. The structure of the ear consists of three parts: the outer ear, the middle ear, and the inner ear; each is designed for a highly specialized purpose. The structure is shown in Fig. 1.8.

The Function of the Outer Ear

The outer ear is the visible part of the ear organ. The outer ear itself consists of two subparts: the pinna and the auditory canal. The folds of cartilage which are attached to the sides of the head, surrounding the ear canal, are called the pinna. Sound waves are reflected and attenuated when they hit the pinna, and these changes provide additional information that will help the brain determine

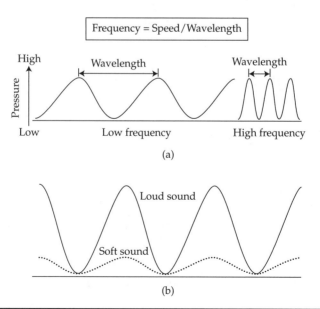

Figure 1.7 Characteristics of a sound wave: (a) wavelength and frequency and (b) amplitude and loudness.

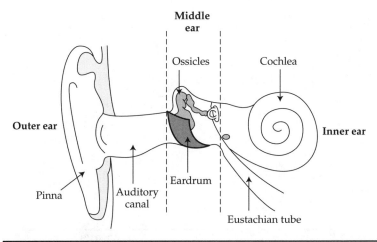

Figure 1.8 The structure of the ear.

the direction from which the sounds came. After hitting the pinna, the sound waves enter the auditory canal, a deceptively simple tube through which sound waves pass to the middle ear. The auditory canal amplifies sound waves that are between 3 and 12 kHz. At the far end of the auditory canal is the eardrum (or tympanic membrane), which marks the beginning of the middle ear.

The Function of the Middle Ear

The middle ear is a small cavity that is connected to the outer ear by the eardrum and to the inner ear by the cochlea. Sound waves traveling through the auditory canal will hit the eardrum. This wave information travels across the air-filled middle ear cavity via a series of delicate bones: the malleus (hammer), the incus (anvil), and the stapes (stirrup). These ossicles, the smallest bones in the body, act as a lever and a teletype, converting the lower-pressure eardrum sound vibrations into higher-pressure sound vibrations at another, smaller, membrane called the oval (or elliptical) window[4] (see Fig. 1.9). Although higher pressure is necessary because the inner ear beyond the oval window contains liquid rather than air, this can sometimes harm the inner ear, especially in cases of sustained loud noises. The middle ear muscles can help protect the inner ear from damage in these cases by reducing the transmission efficiency of the ossicles. These muscles are activated in the presence of such noises, in a preprogrammed response. The middle ear receives information from sounds in wave form which is then converted to nerve impulses in the cochlea.

The Function of the Inner Ear

The inner ear consists of the cochlea and several nonauditory structures. The coiled shape cochlea (see Fig. 1.9) supports a fluid wave

10 Chapter 1

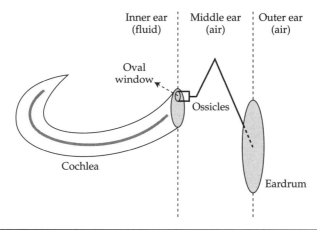

FIGURE 1.9 Schematic anatomy of the ear. Inner components of each part and the way they are connected to each other.

driven by pressure across the basilar membrane, which is attached to the oval window. Vibration of this oval window also causes the basilar membrane to vibrate. The organ of corti is located in a section of the cochlea called the cochlear duct. This organ forms a ribbon of sensory nerves that runs lengthwise down the entire cochlea. The hair cells in the organ of corti, called cilia, are mechanosensors for hearing. When the basilar membrane vibrates, it produces fluid waves in the inner ear which tends to move the cilia cells back and forth. In fact, it is by this movement that the cilia cells transform the fluid waves as a sort of mechanical stimuli into nerve signals.

The Function of the Round Window

In this section, we discuss the function of the round window. The main purpose of the round window is to displace the basilar membrane. However, this membrane is surrounded by an incompressible fluid. In nature, this problem is solved by the round window (see Fig. 1.10). When the oval window is pressed in, some portion of the pliable basilar membrane also bulges. This, in turn, produces a bulge in the round window into the middle ear. Thus, pressure waves are transmitted across the basilar membrane to the round window. Here, the round window acts in a similar manner to a pressure outlet.[1] If there is no round window in the ear, the oval window will not be able to produce any motion.

The Activation of Auditory Afferents

On the basilar membrane, there are a considerable number of hair cells (see Fig. 1.10). The hairs on the hair cells are bent as a result of the bending of the basilar membrane. Ion channels are opened by a filament located between adjacent hairs. By doing this, it allows K^+

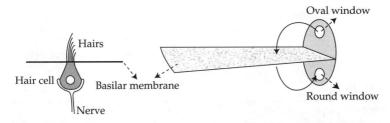

Figure 1.10 The round window plays an important role in the transmission of pressure waves in the basilar membrane.

to enter the hair cell.[1] This, in turn, leads to the depolarization of the hair cells. Finally, the neurotransmitter will be released. This chemical enhances the firing rate in bipolar 8th nerve neurons.

The Pattern of Vibration of the Basilar Membrane

In one study, von Bekesy, made a detailed observation of a traveling wave moving down the basilar membrane near the round window and ending at the helicotrema. In Fig. 1.11 we can see such a wave. Movement of the wave along the basilar membrane is comparable to the movement of a pressure wave along the arterial walls; it is also comparable to the wave that travels along the surface of a pond.[3]

One can easily test this experience using a 4-m-long cord. Attach one of the cord's ends to a solid support and, while firmly grasping the other end, flick your hand down. The wave travels down the cord. Since the material of the cord is uniform, regardless of the location on the cord, we can observe that the size of the wave does not change considerably. However, we know that the basilar membrane properties change. This means that a low frequency causes the wave to increase in size, while a high frequency causes the wave to reduce in size.

The Coding of Frequency of a Sound

Near the oval window, the basilar membrane is narrow and stiff, while at the other end it is wide and floppy. Therefore, for a particular frequency of sound, each portion of the basilar membrane vibrates maximally. High-frequency sounds maximally displace the hair cells near the oval window. On the other hand, low-frequency sounds maximally displace the hair cells at the other end. It can be concluded that there is some sort of place coding of sound frequency occurring on the membrane. In other words, frequency is coded by which neuron is activated, not necessarily by its firing rate.

The Coding of Loudness

Loud sounds vibrate the basilar membrane more than soft sounds. This happens because large vibrations generate higher firing rates in

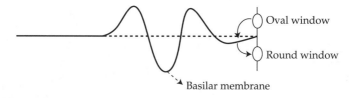

Figure 1.11 Schematics of a wave taken an instant apart.

hair cells. Hence, loudness is encoded by the frequency of APs in a particular hair cell (see Fig. 1.12).

If we take a clarinet as an example of a typical sound we hear on a daily basis, we observe that it produces a complex sound. This is due to the fact that complex sounds have multiple frequencies, which is the distinguishing feature of every musical instrument. The hair cells disseminate this complex sound into its different frequencies. This is the result of the phenomenon that each hair cell encodes the loudness of a particular frequency separately.

Hearing Loss

There are a number of risk factors that can cause failure in the hearing mechanism that can eventually lead to hearing loss. Some of the major causes can be categorized as follows:[1]

Loud sounds: Extremely loud sounds, such as the ones from explosions and gunfire, can rupture the eardrum and break the ossicles, or tear the basilar membrane. Common loud sounds such as lawnmowers or loud music can shear the hairs off the hair cells over a period of time.

Infections: In rare cases, fluid buildup and middle ear infections can rupture the eardrum. Inner ear infections can also damage hair cells.

Drugs: Environmental toxins and some antibiotics can enter hair cells through the open channels and poison them.

Old age: The parts are simply exposed to mechanical wear with time. In old age, blockage in the blood supply can also lead to the death of cells.

1.3 Sense of Smell

The sense of smell, called *olfaction*, involves the detection and perception of chemicals floating in the air. This sense is mediated by specialized sensory cells of the nasal cavity. The chemicals which activate the olfactory system are called odors. Different odors

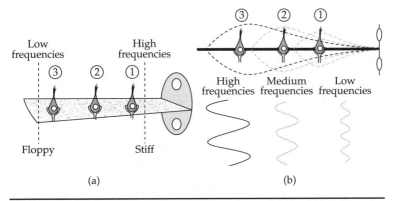

FIGURE 1.12 (a) Basilar membrane frequency characteristic. (b) Loudness is encoded by the frequency of APs.

are attributed to different shapes and sizes of odor molecules that stimulate the olfactory receptor cells. Molecules of odorants passing through the nasal passages dissolve in the mucus lining of the superior portion of the cavity and are detected by olfactory receptors.[4] Mucus is a slippery secretion of the lining of the mucous membranes. It is a viscous colloid containing antiseptic enzymes. It also serves as a trap for small particles such as bacteria and dust, helping to prevent them from entering the body and lungs. Detection of odorant molecules dissolved in mucus by the receptors may occur by diffusion or by the binding of the odorant to odorant binding proteins. Odor receptor nerve cells function like a key-lock system. If the airborne molecules of a certain chemical can fit into the lock, the nerve cell will respond and the brain interprets this response into a smell. The smell of a rose, perfume, freshly baked bread, and cookies are all made possible because of the nose and olfactory system. As they say, the nose knows!

The Olfactory System

Our olfactory system consists of 500 to 1000 different types of smell receptors[5] and, overall, there are approximately 40 million of them in the nasal cavity. They can form an area of up to 16 cm^2 if all of them are placed next to each other. The system is capable of recognizing 4000 to 10,000 different smells.[6] Although this system works efficiently enough in humans, some animals have an even more powerful olfactory system. By comparison, canines have smell cells 100 times larger than humans. A German shepherd dog has about 2 billion olfactory receptors with considerably more receptors per square centimeter. This allows dogs to differentiate among many more types of smells and to be able to sense more diluted amounts of odor present in the environment.

14 Chapter 1

Figure 1.13 represents an anatomy of the human olfactory system. Different segments of this system involved in our sense of smell, including the receptors and their placement in the nasal passage, are also illustrated.

As is evident in Fig. 1.13, olfactory sensory nerves pass through to the olfactory bulb on their way to the brain. The olfactory bulb transmits smell information from the nose to the brain, and is, therefore, necessary for a proper sense of smell. As a neural circuit, the olfactory bulb has one source of sensory input (axons from olfactory receptor neurons) and one output (mitral cell axons). As a result, it is generally assumed that it functions as a filter, as opposed to an associative circuit that has many inputs and many outputs. Its potential functions can be placed into three categories:

- It can enhance discrimination between odors.
- It can enhance sensitivity of odor detection.
- It can filter-out many background odors to enhance the transmission of a few select odors.

Colors include three primary colors, while the rest are combinations of these three; so, too with odors. There are six primary

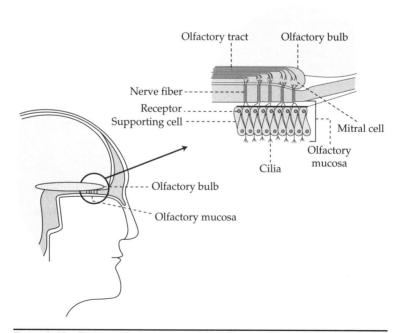

FIGURE **1.13** The graphical anatomy of the olfactory system. (See also color insert.)

(independent) odor qualities: fragrant (scented), ethereal, resinous, spicy, burned, and putrid (rotten). They form a basis to describe any smell sensation as a combination of these six components. The diagram in Fig. 1.14 shows these classifications.

Along with sensing smell, the olfactory system serves several more functions that we use in our everyday behavior:

1. The system acts as a general alarm in cases of fire and smoke.
2. It helps in realizing some components of flavor, in testing food and in everything we eat.
3. We tend to use descriptions of smell by examples in our communications; for instance, we use phrases like "this thing *smells* fishy."

Failure in the ability to sense any of these six main types corresponds to diseases which are called anosmia. These diseases are basically a form of smell blindness. In 2003, two million people in the United States were reported to have some form of anosmia,[6] so it is a rather common dysfunction in the olfactory system. Smell blindness can also happen with age. Some research works suggest our smelling ability increases to reach a plateau at about the age of eight, and declines in old age. Therefore, children are likely to have a more powerful sense of smell than their parents.

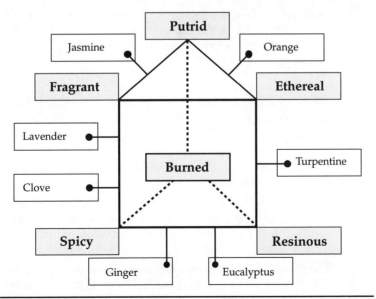

FIGURE 1.14 Six primary odor qualities.

1.4 Sense of Taste

Taste (or, more formally, gustation), refers to the ability to detect the flavor of substances. In humans, this sense complements the less direct sense of smell in the brain's perception of flavor.

The sense of taste is mediated by taste receptor cells. These cells in humans are found on the surface of the tongue and in the pharynx and epiglottis. The tongue, however, is the main organ responsible for detecting taste, as its receptor cells are much more numerous and dense than those of the pharynx and epiglottis.

Taste receptor cells in the tongue are bundled in clusters called taste buds. There are about 10,000 of these taste buds distributed on the surface of the tongue.[7] Taste buds are composed of groups of between 50 and 150 columnar taste receptor cells clustered together like a bunch of bananas. The taste receptor cells within each bud are arranged such that their tips form a small taste pore, and through this pore extend microvilli from the taste cells. The microvilli of the taste cells bear taste receptors.

Interwoven among the taste cells in a taste bud is a network of dendrites of sensory nerves called taste nerves. When taste cells are stimulated by binding chemicals to their receptors, they depolarize. This is then transmitted to the taste nerve fibers, resulting in an action potential that is ultimately transmitted to the brain for interpretation.

Once taste signals are transmitted to the brain, several efferent neural pathways are activated that are important to the digestive function. For example, tasting food is followed rapidly by increased salivation and by low-level secretory activity in the stomach.

Psychophysicists have long suggested the existence of four taste "primaries," referred to as the basic tastes: sweetness, bitterness, sourness, and saltiness. A brief description of each of these basic tastes could be given as follows:

- *Sweetness* is produced by the presence of sugars and a few other substances. Also, chemicals that contain a carbonyl group can generate this taste.

- *Bitterness* is considered to be unpleasant for most people. Denatonium is believed to be one of the bitterest chemicals. To avoid accidental ingestion of harmful chemicals, denatonium is sometimes added to various toxic substances.

- *Sourness* is, in fact, the detection of acidity. The mechanism for detecting sour taste is hydrogen ion channels. In other words, sourness is the detection of the concentration of hydronium ions.

- The presence of sodium or calcium ions leads to *saltiness*. Here, the production of the action potential is due to the sodium ions that have passed through ion channels in the tongue.

These tastes are recognized by the taste buds in different locations of the tongue as shown in Fig. 1.15. The salty/sweet taste buds are located near the front; the sour taste buds line the sides and the bitter taste buds are found at the very back of the tongue.

There are thresholds for activation of taste bundles for detection of each taste. The threshold varies with respect to taste and the chemical substances that lead to it. For instance, the threshold for tasting saltiness (NaCl substance) is 0.01 M (molar) and the threshold for tasting sourness (HCl substance) is 0.0009 M.[3]

Among humans, there is substantial difference in taste sensitivity. In general, girls have more taste buds than boys; this means that girls can recognize a taste with a lower threshold and that they are more capable of differentiating among various tastes than are boys. Some people are supertasters; this means that they are several times more sensitive to bitter tastes and to other tastes than are those that taste normally. Aging can also affect the sense of taste. Babies have taste buds, not only on their tongues, but on the sides and roof of their mouths. This means that babies are very sensitive to different foods. As we grow older, the taste buds begin to disappear from the sides and roof of our mouths, leaving taste buds mostly on our tongues, which also become less sensitive. In addition, harmful external stimuli can also lead to deficiencies in this sense. Heat and aggressive chemicals are the most common samples of these stimuli. Extra hot foods or toxic chemicals entering the mouth can lead to massive receptor cell death and cause a malfunction in the process of identifying various tastes.

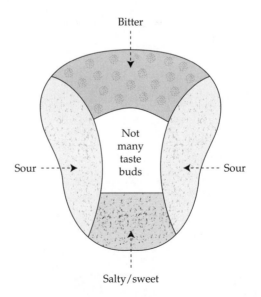

Figure 1.15
The locations in the tongue where the four primary tastes are recognized.

References

1. Tutis Vilis, "The Physiology of the Senses Transformations for Perception and Action," Department of Physiology and Pharmacology, University of Western Ontario, Canada (http://www.physpharm.fmd.uwo.ca/undergrad/sensesweb/).
2. D. Alford and J. Hill, *Excel HSC Biology: With HSC Cards*, Pascal Press, 2004.
3. A. C. Guyton and J. E. Hall, *Textbook of Medical Physiology*, 10th ed., W.B. Saunders, 2000.
4. http://en.wikipedia.org/wiki.
5. I. B. Weiner and D. K. Freedheim, *Handbook of Psychology*, John Wiley and Sons, 2003.
6. http://www.anosmiafoundation.org/suffer.shtml referencing the *The New York Times*, September 22, 2003.
7. A. Claybourne, *The Human Body*, Evans Brothers, 2006.

CHAPTER 2
The Sense of Touch

2.1 Introduction

Touch, also called *tactition*, is the sense by which external objects or forces are perceived through physical contact mainly with the skin. It is our oldest and most primitive sense. Touch is the first sense developed in the womb and the last one we lose before death. In conjunction with sight, our sense of touch contributes to the recognition of the shape and size of objects, enabling us to locate them in the surrounding space. Our sense of touch gives us the feel of the world around us and allows us to tell the difference between rough and smooth, soft and hard, and wet and dry. Figure 2.1 briefly illustrates some common features in our daily lives which would not be possible without this sense.

Apart from its sensory perception, touch has amazing psychological effects on humans and their relations. It is a way of expressing our passion and love to each other. Studies have pointed out that young infants that are affectionately touched by their parents grow more quickly and have a greater immunity to every kind of physiological harm.[1]

As in humans, sense of touch is also of critical importance for some animals. As an example, spiders are not able to see much more than light, dark, and basic shape; but they have a very good sense of touch. Spiders learn more about the world around them by feeling vibrations than by using their eyes. By feeling the differences in various types of vibrations, they can determine what is happening around them. For example, air movement and trapped insects both make the spider's web move, but the spider can differentiate between the sudden, uneven vibrations caused by the insect and the gentle, smooth vibrations caused by the wind. They also use this sense as a means to communicate with each other, by vibrating the web of another spider in a special way.

Unlike the other four senses including sight, hearing, smell, and taste, where the sensory organs are located in specific areas of our body, the sense of touch can be perceived all over. This is because the main sensory organ for touch is our skin, and our skin covers

20 Chapter 2

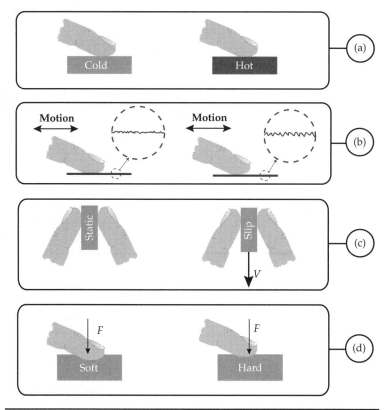

FIGURE 2.1 Some common features that can only be detected by our sense of touch: (a) thermal sense, (b) roughness of surface texture, (c) sense of slip, and (d) pressure and compliance detection. (See also color insert.)

the entire surface of the body. Skin is the body's largest organ and it weighs around 2.8 kilograms. It has about fifty touch receptors for every 100 square millimeters and about five million sensory cells overall.[2] However, these sensory cells are not evenly distributed all over the skin. Some parts of the skin such as fingertips and lips have more nerve endings; these areas are, therefore, more sensitive to touch than are other parts of the body.

Along with its major contribution to the sense of touch, skin simultaneously performs several other functions. Some of its important functions include:

1. *Protection*: Skin is an anatomical barrier, protecting the body from damage and preventing the invasion of microbes and microorganisms.

2. *Heat regulation*: The skin contains a blood supply far greater than its requirements, allowing precise control of energy

loss by radiation, convection, and conduction. Dilated blood vessels increase perfusion and heat loss while constricted vessels greatly reduce cutaneous blood flow and conserve heat.
3. *Control of evaporation*: The skin provides a relatively dry and impermeable barrier to fluid loss, preserving the required amount of fluid within the body.
4. *Communication*: Other people see our skin and can assess our mood, physical state, and attractiveness.
5. *Storage and synthesis*: Skin acts as a storage center for lipids and water, as well as a means of synthesis of vitamin D by action of UV on certain parts of the skin.
6. *Excretion*: Skin eliminates harmful substances resulting from the metabolic activities through the act of sweating.
7. *Immunology*: Skin plays an immunological role, cooperating with Langerhans cells (histiocytosis), as a part of the adaptive immune system.
8. *Sensation*: Skin contains a variety of nerve endings that react to heat and cold, pressure, vibration, and tissue injury, apart from those for touch.

In a sensory system, a sensory receptor is a structure as part of the nervous system that recognizes an internal or external stimulus in the environment of an organism. In response to stimuli, the sensory receptor initiates sensory transduction by creating nerve impulse (action potentials). These action potentials are then transmitted to the brain in the form of neural impulses for interpretation.

Sensory receptors within the human body can be classified by their functional modality for which they react to specific stimulation, as follows:

- *Chemoreceptors* respond to chemical stimuli such as receptors for taste and smell.
- *Mechanoreceptors* respond to mechanical stimuli such as pressure.
- *Nociceptors* respond to damage to body tissues leading to pain perception.
- *Osmoreceptors* respond to the osmolarity of fluids, such as in the hypothalamus.
- *Photoreceptors* respond to light.

In this chapter, we focus on mechanoreceptors and present a classification for these receptors. The human body has two sensory systems which react to contact with external objects. These two sensory systems are the exteroceptive sensor system and the proprioceptive sensor system.

2.2 The Exteroceptive Sensor System

The exteroceptive system reacts to the changes in temperature and the deformation of the skin surface where these quantities are a direct result of contact with external objects. In other words, exteroceptive means that it pertains to a sensory organ which responds to stimuli from outside the body. For the purposes of this book, this includes mechanical stimuli.

Using the described perception of touch in the previous section, touch receptors are types of mechanoreceptors. There are four main types of mechanoreceptors embedded all over the human skin throughout the body, and each is responsible for the reception of specific stimuli; these four include Pacinian corpuscles, Meissner's corpuscles, Merkel's discs, and Ruffini cylinders. Figure 2.2 shows these receptors and their locations within the skin.

Two modes of receptors are: rapidly adapting (RA) receptor or fast adapting (FA) receptor and slowly adapting (SA) receptor. RA will not fire action potentials throughout the duration of a stimulus; rather, it will fire briefly at its beginning and end. It responds to changes in stimulation, but does not continue to respond to constant stimulation. SA is active throughout the period during which the stimulus is in contact with its receptive field and it responds to constant stimulation.

Pacinian corpuscles are nerve endings responsible for sensitivity to deep-pressure touch and high-frequency vibration (250 to 350 Hz).[3] They are located beneath the bottom layer of skin, called

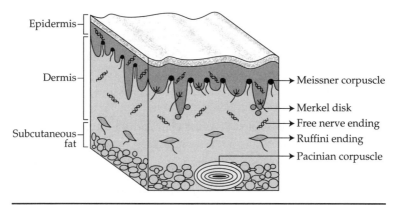

Figure 2.2 Layers of skin and location of its mechanoreceptors. (See also color insert.)

the *dermis* which is subcutaneous fat. They are considered rapidly adaptive receptors. The Pacinian corpuscles are oval shaped and are approximately 1 mm in length and 0.6 mm in diameter.[4]

Meissner's corpuscles are responsible for sensitivity to light touch. As shown in Fig. 2.2, they are located in the top layer of skin, called the *epidermis*. Similar to Pacinian corpuscles, they are rapidly adaptive receptors. Meissner's corpuscles are between 80 and 150 μm in length and between 20 and 40 μm in diameter.[5] They are mostly distributed in the fingertips, the palms, the lips, the face, and the skin of the male and female genitals.

Merkel's discs provide information regarding pressure, vibration, and texture. They are the most sensitive of the four mechanoreceptors to vibrations at low frequencies, at around 15 Hz. They are located in epidermis layer, as shown in Fig. 2.2. In contrast to the two previous receptors, they are slowly adaptive receptors. This means that they have sustained response to stimulation since they are not capsulated, as are the two previous ones.

Ruffini cylinders are sensitive to lateral skin stretch and contribute to the kinesthetic sense and control of finger position and movement. They register mechanical deformation within joints, more specifically angle change, with a specificity of up to 2 degrees, as well as continuous pressure states.[6] They are also believed to be useful for monitoring slippage of objects along the surface of the skin, allowing modulation of grip on an object. Ruffini cylinders are located in the dermis layer.

Figure 2.3 shows a simplified two-dimensional sketch of the skin which is useful for a quick reminder of the locations of each of these receptors. Epidermis has a thickness that varies from 0.05 mm on eyelids to between 0.8 and 1.5 mm on the soles of the feet and the palms of the hands, and the dermis is 2 to 12 times thicker than the epidermis.[7]

Skin hairs also play a part in the sense of touch in addition to their main function of regulating body heat. Hairs are sensitive to touch, acting as levers in which slight changes in position are detected and

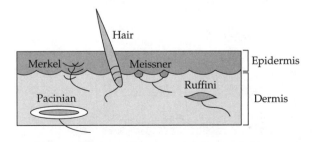

FIGURE 2.3 A two-dimensional sketch of the locations of each of the skin's four receptors.

signaled by a network of nerve endings at the root of each hair[8] as shown in Fig. 2.4.

The Receptive Field

The receptive field of a particular receptor is the area of skin over which tactile stimuli change the activity of the receptor (see Fig. 2.5). Tactile acuity is determined by the ability to discriminate between one and two points of stimulations.[9] Greater tactile acuity is related to a smaller receptive field.

We will use label 1 (punctate) for surface and label 2 (diffuse) for deep. The four receptors in skin are:

- Hair receptors or Meissner is RA1 and surface receptor.
- Merkel is SA1 and surface receptor.
- Ruffini is SA2 and deep receptor.
- Pacinian is RA2 and deep receptor.

Both surface and deep layers contain both RA and SA receptors (see Fig. 2.6).

2.3 The Proprioceptive Sensor System

There are also other types of touch receptors which are mainly located in muscles, muscle tendons, and connective tissue surrounding the joints; however, they are not categorized as mechanoreceptors. In fact, they belong to another sensory system of the body, known as the proprioceptive sensory system.

Proprioception is the sense of the relative position of neighboring parts of the body in static and dynamic situations. Unlike our

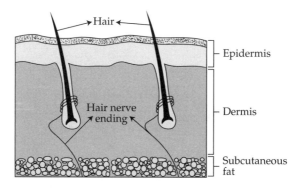

FIGURE 2.4 Structure of hair as a touch sensor and its location within the skin layers.

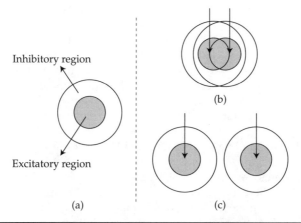

Figure 2.5 (a) The names of regions, (b) pressure applied to two close points feels like one point, and (c) pressure applied to distant points feels distinct.

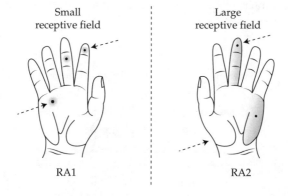

Figure 2.6 Two types of receptive field for RA receptors (RA1 and RA2).

traditional five senses (taste, smell, touch, hearing, and vision) by which we perceive the outside world, proprioception is a distinct sensory modality that provides feedback solely on the status of the body internally. It is the sense that indicates whether the body is moving with required effort, as well as where the various parts of the body are located in relation to each other.

So far, four groups of sensors have been identified which could play a role in proprioception. Two groups of these sensors are called *annulospiral* and *flower spray endings*.[8] They are situated in muscle spindles and it is believed that they respond to passive stretching of muscles. Figure 2.7 shows these sensors schematically. The body's posture is maintained using the assistance of these types of nerve endings.[8]

Another group of nerve endings, Golgi tendon organs found where the tendon joins with the muscle, is responsible for sensing the total muscle tension. This third group is situated in muscle tendons (see Fig. 2.8) and responds to muscle tension generated. This tension can be produced in two ways. The first source could be an external or passive force and the second one could be the muscle itself actively causing the tension.[8]

The fourth group of nerve endings is sensory receptors in and around joints, located primarily in the ligaments which stabilize the joint.[10] Limb joint angles are measured by this group. The detection of small angular movements of joints is achieved using this type of nerve endings. As an example, if one considers the shoulder joint movement, these nerve endings can discern the displacements of about half a degree occurring in less than two seconds.[8]

Now that the concept of touch receptors, including their structure, abilities, and functions has become clear to us, it is time to explore the next step; this step explains how mechanical stimuli features are electrically encoded by the skin's touch receptors to make them interpretable by our brain and to make the brain react in an appropriate way.

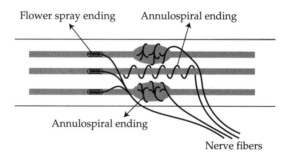

Figure 2.7 Annulospiral and flower spray endings located in muscle spindle.

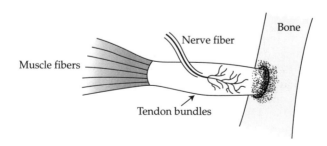

Figure 2.8 Nerve endings located in muscle tendons.

2.4 Transduction of Mechanical Stimuli to Neural Impulses

The process of touch originates with a physical contact between the touch sensory organ such as the skin and an external stimulus. As a result of this physical contact, the mechanical stimulus such as pressure deforms the mechanoreceptor's membrane (see Fig. 2.9, related to Pacinian receptor).

Next, the ion channels of mechanoreceptor's sensory cells become open, letting Na$^+$ and K$^+$ ions pass through their membrane. The ions are transported in such a way that the Na$^+$ ions go into the cell while the K$^+$ ions flow out of the cell. By this series of events, the receptor cell depolarizes. The process is shown in Fig. 2.10.

After the cell becomes depolarized, action potentials are generated and propagated down the axon of the cell. Most touch afferents have myelin-coated axons in which action potentials hop from gap to gap, thus speeding up the action potential propagation process. The concept is illustrated in Fig. 2.11. Anatomical components of a sensory cell are also shown in Fig. 2.12.

Our brain determines the magnitude of sensed stimulus by the frequency of neural impulses produced by mechanoreceptory

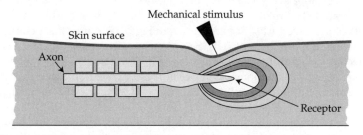

FIGURE 2.9 The first step of the transformation process. The mechanical stimulus is deforming the receptor.

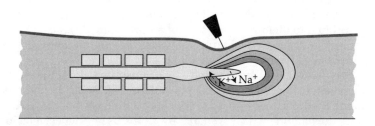

FIGURE 2.10 Receptor depolarization in the second step of the process, as a result of first step.

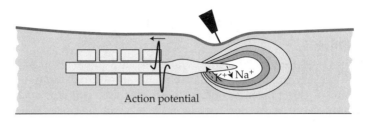

FIGURE 2.11 Generation of action potential and its propagation down the axon. (See also color insert.)

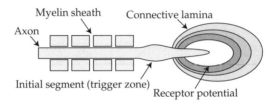

FIGURE 2.12 Schematic of anatomical components of a sensory cell.

cells; the greater the stimulus, the more the receptor depolarizes, causing more sets of action potentials-per-second (APs/Sec) to be generated. The relation between magnitude of the stimulus and the sets of produced APs/Sec are shown in Fig. 2.13. A graphical relation between the numbers of APs/Sec versus pressure, as an example of mechanical stimulus, is also presented in Fig. 2.14. The graph shows that the typical relationship between these two is nonlinear and the receptor cell tends to enter to its saturated state to saturate at high amount of stimulus.

It is by this mechanism that the neural impulses are being transducted from mechanical stimuli and sent to the brain. Sometimes, however, different stimuli could result in the same response by the receptors. That is, the APs/Sec produced by Merkel's discs receptors in response to a low-frequency vibration sensed by them, could be similar to the ones produced by Meissner's corpuscles in response to a light pressure. This is illustrated in Fig. 2.15. Therefore, it is important for the brain to identify which type of receptors are producing these impulses in order to distinguish between the stimuli. How is this problem solved in our body's tactile sensory system?

The Internet is an example that illustrates how the same phenomenon could happen in the outside world. While the Internet is being used, the message we send, along with those of other users, moves down a line which is shared by all the users. Hence, a tag or label is given to each packet of information in order to distinguish our message from that of others. Finally, a decoder is used at the

The Sense of Touch 29

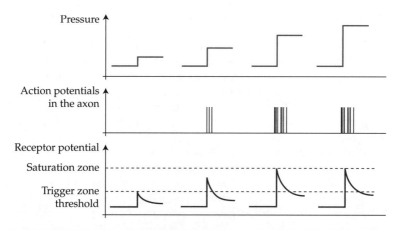

FIGURE 2.13 Effect of magnitude of mechanical stimulus on the action potential produced by the receptor cell.

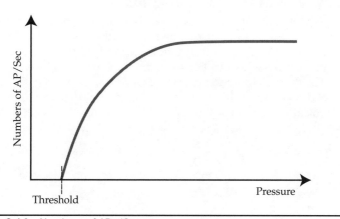

FIGURE 2.14 Numbers of APs/Sec versus pressure.

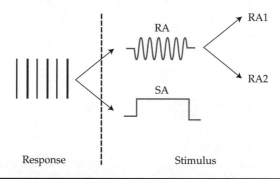

FIGURE 2.15 Different stimuli can result in the same response.

end of the line so that it can separate our packet from that of others (Fig. 2.16a).

Our body has solved this same problem in a different way. Each type of touch sensors has its own private line in body. These routes are called the *labeled lines* and are shown in Fig. 2.16b. As a result of these separate pathways, there is no reason for encoding and decoding each packet of information so the brain can exactly distinguish between the produced impulses by the different mechanoreceptors, even if the impulses occur at the same.

With the understanding of the process of generating neural impulses in response to mechanical stimuli by the respective mechanoreceptors, there is now just one final step. In order to complete the puzzle and to understand the function of our sense of touch, we must now investigate how, and to where, the impulses are transferred.

2.5 Pathways of Tactile Information

The sensory information produced by receptory cells, in the form of neural electrical impulses generated by the foresaid mechanism from all over the skin on different parts of body, is conveyed to the central nervous system by afferent neurons. These neurons vary in their size, structure, and properties. Generally, there is a correlation between the type of sensory modality detected and the type of afferent neuron involved. So, for example, faster, thicker, and myelinated neurons conduct pain for a faster and more

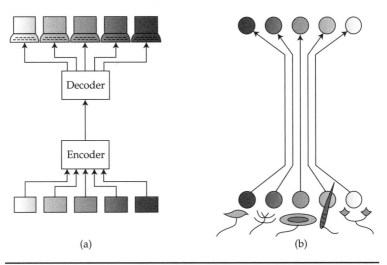

FIGURE 2.16 (a) The mechanism which separates messages in the same line and (b) the solution applied in our body's tactile sensory system.

efficient transition. The afferent neurons transfer tactile data from the surface of the skin through the spinal cord to the brain. The pathways of different parts of the body to the spinal cord are shown in Fig. 2.17.

The spinal cord includes ascending pathways from the body to the brain. The major target is the somatosensory cortex, as seen in Fig. 2.18. Areas of this part of the human brain map to certain areas of the body, depending upon the amount or importance of sensory information from that area. For example, there is a large area of cortex devoted to sensation in the hands, while the back has a much smaller area. After receiving and processing the information, commands from the brain are sent to organs by another area of the brain called *the primary motor cortex*, shown in Fig. 2.18.

The diagram in Fig. 2.19 shows the typical pathway for tactile information, from its inception all the way to the brain. For example, in sharp pain reflex, the limb is pulled towards the body out of harm's way, as shown in the dash line in Fig. 2.19, and normal pain information continues to the brain for more considered action.

2.6   Special Features of Tactile Sensing

In Fig. 2.20, we see the top view of two rectangular cubes with the same material, appearance, and dimensions. We can detect the embedded object; for example, we detect the hole in two rectangular cubes only with the sense of touch and with no other senses.

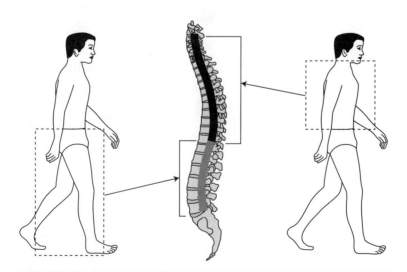

Figure 2.17   Pathways of afferent neurons from different parts of body to the spinal cord.

32 Chapter 2

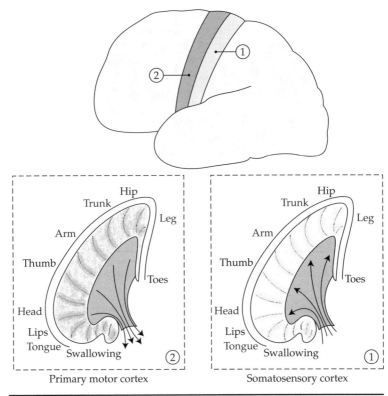

FIGURE 2.18 Areas of the brain involved in receiving tactile information from, and sending appropriate commands to, the body's different parts.

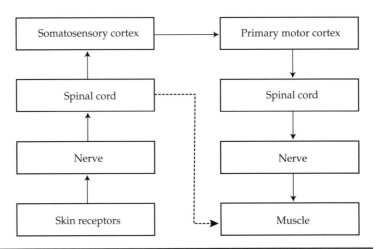

FIGURE 2.19 Tactile information pathway.

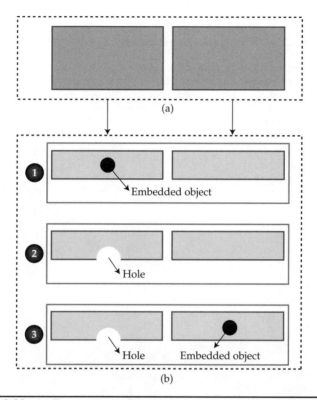

FIGURE 2.20 (a) The top views of two rectangular cubes which are similar visually and (b) the front views of three cross sections of possible situations for them.

We now have completed our puzzle in understanding the sense of touch. Its definition, physiology, components, and functions were all briefly explored in this chapter. A more engineering approach will be employed in the coming chapter. We will be introduced to some engineering definitions about this sense, while trying to formulate its function; we will first do so in a simple manner, and will then move to more complicated and accurate domains, in order to simulate this sense and to take advantage of its features through engineering techniques.

References

1. P. Mccormick, *Just the Right Touch*, Claretian Publications, 1999.
2. D. Kuhl, *What Dying People Want: Practical Wisdom for the End of Life*, Perseus Books Group, 2003.
3. G. Mather, *Foundations of Perception*, Psychology Press, 2006.
4. W. S. Hoar, *General and Comparative Physiology*, Prentice-Hall, 1966.

5. V. B. Mountcastle *The Sensory Hand: Neural Mechanisms of Somatic Sensation*, Harvard University Press, 2005.
6. N. Hamilton, *Kinesiology: Scientific Basis of Human Motion*, McGraw-Hill, 2008.
7. J. Bensouilah, Ph. Buck, R. Tisserand, A. Avis, *Aromadermatology: Aromatherapy in the Treatment and Care of Common Skin Conditions*, Radcliffe, 2007.
8. R. A. Russell, *Robot Tactile Sensing*, Prentice-Hall, December 1990, 350 pages.
9. Ph. J. Corr, *Understanding Biological Psychology*, Blackwell, 2006.
10. T. A. McMahon, *Muscles, Reflexes, and Locomotion*, Princeton University Press, 1984.

CHAPTER 3
Introduction to Tactile Sensing and Tactile Sensors

3.1 Tactile Sensing

As described in Chap. 2, our tactual perception of the outside world is mainly based upon information provided by the body's two types of sensing systems: proprioceptive sensing and exteroceptive sensing. Proprioceptive sensing, also called *kinesthetics*, detects body position, weight, or movement of the muscles, tendons, and joints. On the other hand, exteroceptive sensing, also called *cutaneous sensing*, refers to all sensations including tactile, thermal, and pain.

Tactile sensation is the process of determining physical properties and events of an object through physical contact with that object. A tactile sensing system, therefore, is a system that is capable of detecting and collecting such information by means of tactile sensors. In other words, it is the process of detecting and measuring the spatial distribution of forces on a sensory area. Unlike other sensing modalities in humans, tactile sensing is determinant and direct. It is not distorted by perspective, confused by external lighting, or greatly affected by the material constitution or surface finish of objects. In addition to the direct information perceived by tactile sensing, it also provides us with information to maintain the posture of our body, to monitor walking, to grasp items, to manipulate objects, and to warn of physical danger.

Using another type of classification, tactile sensing can be divided into *active sensing* and *passive sensing*. Passive sensing, or *static tactile sensing* (STS), is concerned with analyzing static tactile data produced by the cutaneous sensory network. Static tactile data refers to information that is perceived by a static contact with an object. In active sensing or *dynamic tactile sensing* (DTS), motion is actively used to extract more information. This type of sensing integrates the cutaneous sensory information with kinesthetic data; this includes information that is perceived by moving the skin on the surface of an object. These two types both employ the tactile sensing concept, but do so in different functional ways.

For a better understanding of static and dynamic tactile sensing, let us take an example. If you place your hand against a textured surface such as wood or cloth and hold it still, it is very difficult to identify the texture. The information that is perceived in this way is limited to items such as the surface temperature and its hardness or softness. However, as soon as you begin to explore the surface by moving your fingers lightly over it, you can describe it in more detail; with this knowledge you can evaluate the roughness of its surface texture, its shape and size and other useful information.

This information, in addition to that acquired by proprioceptive and cutaneous sensing, is acquired utilizing active and passive tactile sensing in our overall haptic perception. *Haptics* describes the general sense of touch which incorporates all kinesthetics, cutaneous, static, and dynamic tactile sensory information. Tactile sensing conveys information which helps the brain to determine a complete picture of the spatial, geometrical, and physical properties of objects and sensations that surround us. This is called *haptic perception* and includes thermal sensation, pain sensation, and the kinesthetic senses. Figure 3.1 shows how these sensations are incorporated in the haptic perception described above. As is evident in Fig. 3.1, by moving our hand on the surface of an object, we can determine its surface texture; by touching the surface with our fingertips, we can perceive its hardness and estimate its temperature; when holding an unsupported object, we can estimate its weight and determine its shape and volume.

The process during which all of the above information is produced, conveyed to the brain, and perceived was detailed in Chap. 2. For a

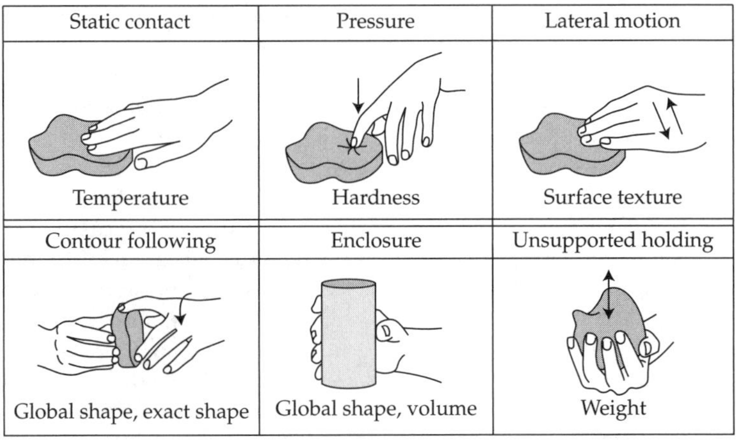

Static contact	Pressure	Lateral motion
Temperature	Hardness	Surface texture
Contour following	Enclosure	Unsupported holding
Global shape, exact shape	Global shape, volume	Weight

Figure 3.1 Data extraction using haptic perception.

quick reminder, the diagram in Fig. 3.2 briefly illustrates the steps of this process. When we touch an object, tactile receptors become stimulated. Raw tactile data is then generated and transmitted through the nervous pathways to the brain for processing. After the data is processed, properties of the object are perceived such as temperature and roughness of the surface.

Contrary to other human sensory systems, such as visual systems and auditory systems, progress in the artificial concept of mimicking tactile sensing has been slow. There are three main reasons for this and these are categorized as follows:[1]

First, unlike the other four sensing modalities, tactile sensing does not possess any localized sensory organ. In fact, the sense of touch operates all over the skin as a distributed process. Additionally, in terms of the area covered by the senses, the transduction of tactile signals is distributed over a considerably wider area than in a single localized sensory organ. Organs such as the eyes and ears belong to the latter category. As a result, the simulation of this phenomenon through the creation of an artificial tactile device is a much more difficult task than is developing a discrete artificial sensing organ.

Secondly, tactile sensing is complex by nature and it is not an easy task to find suitable technological analogies in science and engineering for its functions. In other words, tactile sensing through the skin is not a simple transduction of one physical property into an electronic signal. This is because touch includes the detection of softness, stiffness, shape, texture, force, size, temperature, and other related physical properties. So far, we have not found out exactly how these different aspects of the tactile phenomena are related. Also, more research is required to understand how they are processed by the nervous system.

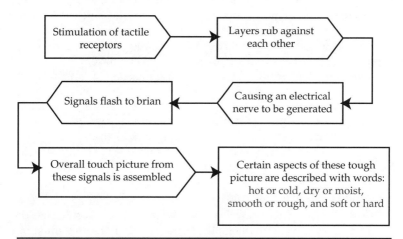

Figure 3.2 The process of tactile sensing.

Third of all, while we know that sight and sound have well defined physical quantities, we do not know the best measures to adopt for a tactile sensor. For example, the intensity of what a hearing organ should measure is physically quantified in decibel (dB) units; similarly, the intensity of what a visionary organ should see based on the intensity of the light is physically quantified in candelas. Tactile sensing, however, is much more complicated and difficult to physically quantify. A large range of physical properties can be transduced and used as tactile signals. However, at this stage, we do not clearly know which is the most appropriate for applications and which should be developed further.

All of the above problems are related to sensors as the means to gather tactile information. However, in order for progress to occur in any of the above areas, a clear understanding of tactile sensors, their definition, functions, capabilities, and limitations is required. The next section will define tactile sensors and provide a brief description of their functions.

3.2 Tactile Sensors

A sensor is a device that measures a physical quantity and converts it into a signal which can be read by an observer or an instrument.[2] A *tactile sensor* is a transducer which is capable of detecting and measuring a given property of an object by making physical contact with it and converting that information into a form which can be processed. Its main application is in artificial tactile sensing where it serves as a data acquisition device.

Artificial tactile sensing is a system that, by means of engineering devices and techniques, aims to simulate the exact tasks of human tactile sensing. In such a system, information is acquired by the sensor and converted into a form compatible with the processor; by processing data and extracting the properties of touched objects, this new form then attempts to simulate the function of the brain. The six broad groupings of carriers of such information are: radiant, mechanical, thermal, electrical, magnetic, and chemical.

One good example in the possible areas of application for tactile sensors is the robot. A robot can be considered an information-processing system where the tactile sensor is mounted on its gripper. The processor is a computer which controls the robot and the actuators provide the system output to the servomotors. Based on information received from the tactile sensor, this system could control the motion of the robot.[3]

Terminology of Artificial Sensors

In order to properly describe the function of a tactile sensor mathematically, some mathematical and physical features must be defined so that they can quantify the sensor's characteristics. These

Introduction to Tactile Sensing and Tactile Sensors

features include: resolution, transfer function, sensitivity, calibration, linearity, hysteresis, accuracy, span, noise, repeatability, reliability, and response time. These features will each be defined and described.

Resolution

Resolution is defined as the minimum increments of stimulus which can be sensed.[4,5] In other words, this specification is the smallest detectable incremental change of input parameter that can be detected in the output signal. As an example, we will look at a tactile sensor which can detect forces within the range of 0 to 20 N, with a resolution of 0.2 N. This means that the sensor is capable of sensing forces, or sensing changes in the exerted force, as small as 0.2 N. This sensor cannot detect an exerted force that varies as little as 0.1 N, so its output signal will remain unchanged.

Transfer Function

The functional relationship between the physical input signal and the electrical output signal is represented by a *transfer function*.[6] A transfer function represents the input-to-output relationship in an $S = f(X)$ form, in which X is the input, S is the output, and f is the transfer function. Usually, this relationship is represented as a graph, showing the relationship between the input and output signal; the details of this relationship may constitute a complete description of the sensor characteristics.[3,6]

There are two possibilities for this function. First of all, it may have the form of a simple linear connection. Secondly, it might have a nonlinear dependency, such as a logarithmic, exponential, or power relation. In many cases, the relationship is unidimensional, meaning that we have the output versus one input stimulus. A unidimensional linear relationship is represented by Eq. (3.1):

$$S = a + bX \quad (3.1)$$

where a is the intercept or the output signal at zero input signal and b is the slope, which is sometimes called sensitivity.[5]

The sensitivity b is not a fixed number for a nonlinear transfer function. At any particular input value, X_0, it can be defined as shown in Eq. (3.2):

$$b = \frac{dS}{dX}\bigg|_{X=X_0} \quad (3.2)$$

Sensitivity

Sensitivity is described in terms of the relationship between the physical signal (input) and the electrical signal (output) and is generally the ratio between a small change in the input to a small change in the output signal.[6] As such, it may be stated as the slope of the transfer function.

Calibration

For a specific sensor, the relationship between the input and output signal is defined as the *calibration of the sensor*.[7] Typically, for the calibrating of a sensor, a known physical input is provided to the system and the output is recorded. The data is plotted on a curve called a *calibration curve* as shown in Fig. 3.3. For values of an input less than X_I, the sensor has a linear response; for values greater than X_I, the calibration curve becomes nonlinear and less sensitive until it reaches a saturated value S_S of the output signal.

Linearity

When the relationship between input and output of a sensor follows a straight line such as when the output is proportional to the input, it is said that the sensor behaves *linearly*. For linear systems, it is easier to formulate a relation between the input and output, and it is also easier to calibrate the system. When dealing with biological tissues, nonlinearity is mostly related to the viscoelastic nature of the human sensory system.[8]

Hysteresis

Some sensors do not return to the same output value when the input stimulus returns to its initial value in a path different from the one it had previously approached. This is more readable in cyclic inputs,

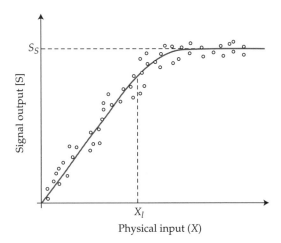

Figure 3.3 A typical calibration curve.

when the input reaches the same value twice, both on its way and on its return. Regarding the output signal of a transducer, *hysteresis* is the variation between upscale and downscale approaches to the same position.[9] Figure 3.4 shows this difference graphically. Take a look at point x_1 on the graph. The sensor steadily indicates an increasing output (moving upscale) as it crosses through position x_1. It then reverses direction and steadily indicates a decreasing reading (moving downscale); it passes again through position x_1, and then to the zero position. There is a slight difference in the reading recorded for the increasing and decreasing approaches to position x_1. This difference is due to hysteresis. Hysteresis mainly occurs due to viscoelasticity in the sensor material and the presence of frictional forces on the surface of contact between the sensor and the object.[5]

Accuracy

Accuracy is measured as the largest expected error between actual and ideal output signals; for instance, this would be the difference between the value which is computed from the output voltage and the true input value.[5,6] For example, consider a linear displacement sensor that should ideally generate 2 mV per 1 mm displacement; that is, its transfer function is linear with a slope of $b = 2$ mV/mm. However, in the experiment, a displacement of $X_{true} = 15$ mm produced an output of $S_{exp.} = 30.4$ mV. Converting this number into the displacement value by using the inversed transfer function, we

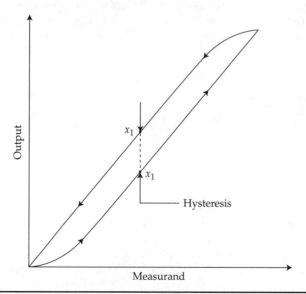

FIGURE 3.4 Hysteresis error in terms of the measured quantity.

would calculate that the displacement was $X_{exp.} = S_{exp.}/b = 15.2$ mm; that is $X_{exp.} - X_{true} = 0.2$ mm more than the true value. This extra 0.2 mm is an error. Therefore, in a 15 mm range, the sensor's absolute accuracy is 0.2 mm or, in relative terms, the accuracy is (0.2 mm/15 mm) × 100% = 1.3%. If we repeat this experiment many times without any random error, and every time we observe an error of 0.2 mm, we may say that the sensor has a systematic accuracy of 0.2 mm over a 15 mm dynamic range.[5]

Span or Dynamic Range

There are two definitions of *span*, which lead to the same concept. The range of input physical signals (stimuli), for which the sensor is capable of converting into electrical signals, is one definition of span or dynamic range.[5,6] Signals outside this range are expected to cause unacceptably large inaccuracy. Span is also called *input full scale (FS)*.[5] Sometimes, span is also defined as the ratio of the maximum value of the physical input signal to the minimum value of the physical input signal that can be converted to electrical signals by the sensor. In other cases, it might be defined as the highest possible input value that can be applied to the sensor without causing an unacceptably large inaccuracy. For sensors with a very broad and nonlinear response characteristic, a dynamic range of the input stimuli is often expressed in decibels, which is a logarithmic measure of ratios of either power or force (voltage). Equations (3.3) and (3.4) represent this ratio for force and power, respectively.[4] It should be emphasized that decibels do not measure absolute values, but only a ratio of values. A decibel scale represents signal magnitudes by much smaller numbers which, in many cases, is far more convenient.

$$1 \text{ dB} = 20 \times \log \frac{X_2}{X_1} \tag{3.3}$$

$$1 \text{ dB} = 10 \times \log \frac{P_2}{P_1} \tag{3.4}$$

Noise

Noise includes the undesirable signals that occur in the sensor output not caused by the sensor input.[3] Inherent noise and interference or transmitted noise are two basic classifications of noise for a given circuit.[5] The noise arising within the circuit is *inherent noise* and the noise picked up from outside the circuit is *interference noise*. Both categories may present a substantial source of errors and should be kept to a minimum.

Repeatability

When the sensor is applied over a set of conditions, and then identical conditions are met again, the difference between these consecutive values is called *repeatability* or reproducibility.[9] This is usually tested by maintaining the conditions and then applying the sensor by changing the input between fixed points.

Reliability

Reliability is the ability of a sensor to perform a required function under stated conditions for a stated period of time.[5] In statistical terms, it is expressed as a probability that the device will function without failure over a specified time or a number of uses. It contains two concepts.[9] The first one is *intrinsic reliability*, which is a function of the quality of design and indicates the reliability that is theoretically possible. The second concept is *achieved reliability*, which is the operational reliability of the sensor.

Response Time

Response time, shown as T_r in Fig. 3.5a, is the amount of time that elapses between the application of an instantaneous change in the physical input to the sensor and the resulting indication of that change in the sensor output.[9] In other words, the output state of a transducer does not change instantly when the input signal changes. There is another parameter relating to this concept, called *decay time*. Decay time, shown as T_d in Fig. 3.5b, is the required time in response to a negative going step-function change of the input parameter.[10] T_{on} and T_{off} in Fig. 3.5 are the onset and the offset times, respectively, of

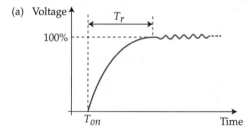
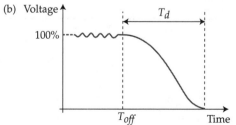

FIGURE 3.5 (a) Definition of response time and (b) definition of decay time.

the input signal. For an ideal tactile sensor, a fast response time is normally considered to be less than 25 ms.[11] In the case of human tactile receptors, the response time is about 1 ms.[12]

Other importance features for designing a tactile sensor are: temperature variation, stability, mass, size, power consumption, presence of over-force protection, cost, and robustness.

Some of the above mathematical features are also applicable to human tactile sensing. The values of these features for fingertips[12] are summarized in Table 3.1.

In order for a tactile sensor to be efficient enough for use in an artificial tactile sensory system, a criterion must be met depending upon the area of application; this criterion should include the standard values for some of the features mentioned above. Harmon was a scientist who analyzed the perceived need for tactile sensing and the specification of tactile devices in various robotic applications; he proposed a set of requirements, shown in Table 3.2, as desirable performance specifications.[13]

Comparing the data in Tables 3.1 and 3.2, it can be inferred that, in addition to their physiological complexities and state-of-the-art functions, the fingertips are also ideal tactile sensors from an engineering point of view.

Some Other Specifications for Tactile Sensors

In addition to the described mathematical features of tactile sensors, there are also some specifications that are not mathematically described; yet, they are qualitative and should also be considered when designing these sensors. These include immunity to temperature variations, size and weight, power consumption, and durability. All are of great importance from the functional point of view and should be taken into consideration according to the conditions under which they will be applied.

Classification of Tactile Sensors

Human tactile sensory systems are divided into STS and DTS. In like manner, sensors are classified as being either static or dynamic.

Parameter	Value
Frequency response	0–100 Hz
Response range	0–100 g/mm^2
Sensitivity	≈0.2 g/mm^2
Resolution	1.8 mm

TABLE 3.1 Sensory Specifications of the Human Fingertip.

Introduction to Tactile Sensing and Tactile Sensors 45

Harmon's design criteria	Character
Sensing surface	Compliant and durable
Spatial resolution between sensing points	1 to 2 mm
Number of sensing points in an array	Between 50 and 200
Minimum force sensitivity	1 gram force (0.01 N)
Dynamic range	About 1000:1
Output response	Monotonic, not necessarily linear
Frequency response	At least 100 Hz
Stability and repeatability	Good
Hysteresis	Low

Source: M. H. Lee, "Tactile Sensing: New Directions, New Challenges," *International Journal of Robotics Research*, vol. 19, no. 7, 2000, pp. 636–643. Copyright © 2000 by Sage Publications. Reprinted by Permission of SAGE.

TABLE 3.2 Harmon's Design Criteria for Tactile Devices in Robotic Applications.

Static tactile sensors, also called *intrinsic sensors*, measure pressure and forces by holding on to the object with a grasping mechanism. *Dynamic tactile sensors*, also called *extrinsic sensors*, measure the characteristics of the touched object during motion on its surface. A typical sample of these sensors is shown in Fig. 3.6.

The main differences between the data perceived by static and dynamic sensors include:[14]

1. *Derivative information*: Various maneuvers need information that is the time derivative of static sensor data. We should employ dynamic sensors for these maneuvers because these sensors can measure quantities by moving on object over time. As a result, we can measure data with respect to time.

2. *Surface scanning*: If we want to study and sample a certain surface without using any motion, then we will need a number of static sensors placed in certain locations on the surface. On the other hand, if we use a single dynamic sensor for sampling a surface, we will be able to scan it more easily. Therefore, in terms of capability, surface scanning is more efficient when using dynamic sensors.

3. *Measuring mechanical properties*: By moving sensors, it is possible to generate shear stresses. Surface roughness and textures can be obtained through this approach. It should be

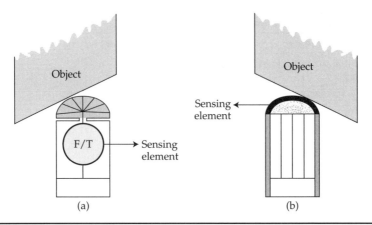

Figure 3.6 (a) A schematic of an intrinsic sensor. F/T means force/torque. Note how the sensing element is placed and (b) a schematic of an extrinsic sensor. The sensing element is spread over the sensor, unlike in the intrinsic type.

noted, however, that when there is no movement, this sort of stress does not exist.

4. *Detection of surface changes:* One method of obtaining information about surface texture is to move the sensor on a surface.

5. *Friction:* Movement is also important for analyzing the frictional properties of a surface. By doing so, the friction coefficient of the surface can be measured.

6. *Relative measurements:* Using a single sensor moving quickly on a surface, we can detect the small differences in adjacent surfaces. Therefore, we will need fewer sensors, which is a great advantage.

7. *Mechanical transients:* Dynamic sensors are able to detect the mechanical transients that are associated with the vibrations produced, such as by grasping objects.

This chapter has described the definition, classifications, and characteristics of tactile sensory systems and sensors. We have now become familiar with the concept of tactile sensing and the function of tactile sensors. In Chap. 4, we will explore the different structures and technologies of common tactile sensors which are now widely used in industry, and we will examine their working concepts based on their engineering characteristics.

References

1. M. H. Lee and H. R. Nicholls, "Tactile Sensing for Mechatronics—A State of the Art Survey," *Mechatronics*, vol. 9, no. 1, 1999, pp. 1–31.
2. http://en.wikipedia.org/wiki/sensor
3. R. A. Russell, *Robot Tactile Sensing*, Prentice-Hall, December 1990.
4. R. Perez, *Design of Medical Electronic Devices*, Academic Press, 2001.
5. J. Fraden, *Handbook of Modern Sensors Physics, Designs, and Applications*, 3rd ed., Springer, 2004.
6. J. S. Wilson, *Sensor Technology Handbook*, Elsevier Inc., 2005.
7. J. G. Webster, *The Measurement, Instrumentation, and Sensors Handbook*, CRC Press, 1999.
8. H. Singh, "Design, Finite Element and Experimental Analysis of Piezoelectric Tactile Sensors for Endoscopic Graspers," M.A.Sc. dissertation, Concordia University, Canada, 2004. (Dissertations & Theses: A&I database, Publication No. AAT MQ94733.)
9. D. S. Nyce, *Linear Position Sensors Theory and Application*, John Wiley & Sons, Inc., 2004.
10. J. J. Carr, *Sensors and Circuits: Sensors, Transducers, and Supporting Circuits for Electronic Instrumentation, Measurement, and Control*, Prentice Hall, 1993.
11. L. Wang, "Microfabricated Shear-Sensitive Tactile Sensor: Development and Application," Ph.D. dissertation, University of Illinois at Urbana-Champaign, Illinois, 2001. (Dissertations & Theses: A&I database, Publication No. AAT 3030489.)
12. J. Dargahi and S. Najarian, "Human Tactile Perception as a Standard for Artificial Tactile Sensing—a Review," *International Journal of Medical Robotics and Computer Assisted Surgery*, vol. 1, no. 1, 2004, pp. 23–35.
13. M. H. Lee, "Tactile Sensing: New Directions, New Challenges," *International Journal of Robotics Research*, vol. 19, no. 7, 2000, pp. 636–643.
14. R. D. Howe, "Dynamic Tactile Sensing," Ph.D. dissertation, Stanford University, California, 2001. (Dissertations & Theses: A&I database, 1991 Publication No. AAT 9115788.)

CHAPTER 4
Introduction to Tactile Sensing Technologies

4.1 Introduction

Technologies used within tactile sensing are mostly categorized by the methods which are employed for the transduction process. In Chap. 3, transduction was defined as converting one form of energy or message into another. This is what happens in the real physiology of human sensing organs as they transduce physical energy into a nervous signal.

The technologies used for transduction in tactile sensory systems are mainly based on detecting and measuring changes in resistance, capacitance, voltage generation, inductance, and optical properties. Depending on their versatility and suitability for specific industrial functions and conditions, these technologies are chosen as the transduction method.

There are different types of sensors used according to the respective transduction technology. The most common types include capacitive, magnetic, inductive, conductive elastomeric, optical, piezoresistive, and piezoelectric type sensors. These types are discussed in the following sections.

4.2 Capacitive Sensors

This type of tactile sensor deals with measuring the capacitance which varies under an applied load. The capacitance of a parallel-plate capacitor depends on the distance between the plates, permittivity of dielectric materials, and their areas. These sensors also contain an elastomer separator between the plates that provides compliance to capacitance according to the applied load. These sensors are capable of detecting touch by sensing the applied normal or tangential forces on them; however, they are not efficient enough for distinctions between these two types of forces.

The structure of a single tactile sensor unit (also called a *tactile cell*) of this type basically consists of three layers. These include two plates made from a conductive material and a dielectric layer in between, which is usually silicone or air. For this sensor to measure the applied force, one plate of the capacitor actually acts as a diaphragm; when force is applied to it, the distance between the two plates decreases, increasing the capacitance. The inverse gap relationship is highly nonlinear and the sensitivity drops significantly with larger gaps. The basic diagram of the structure of a capacitive sensor is shown in Fig. 4.1. The change in capacity is eventually converted into a change in voltage or frequency by using an appropriate circuit.

Let A be the area of the plates. Assuming the distance d between the top and bottom plates is much smaller than the plate dimensions, the capacitance of the cell can be expressed by Eq. (4.1):

$$C = \varepsilon_0 \varepsilon_r \frac{A}{d} \qquad (4.1)$$

where $\varepsilon_0 = 8.85 \times 10^{-12}$ F·m^{-1} is the permittivity and ε_r is the dielectric constant of the dielectric layer.

When an external force is applied to the tactile cell, d is reduced and, therefore, the capacitance is increased. By measuring this capacitance, the tactile information is perceived. Normal force changes the distance between the plates while tangential force changes the effective area between the plates as demonstrated in Fig. 4.2.

For an array of capacitive sensors, a number of tactile cells are connected together to produce a single output. This is called *multiplexing*. In electronics, telecommunications, and computer networks, the term multiplexing is used to refer to a process where multiple analog signals or digital data are combined into one signal. The device that performs the multiplexing is called a *multiplexor* (*MUX*). In sensor technology, one method of multiplexing is to construct each tactile cell individually and use independent multiplexing, as seen in Fig. 4.3.

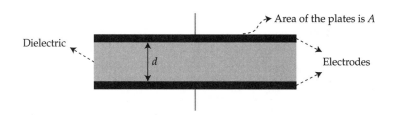

Figure 4.1 A schematic of the structure of capacitive sensor.

Introduction to Tactile Sensing Technologies

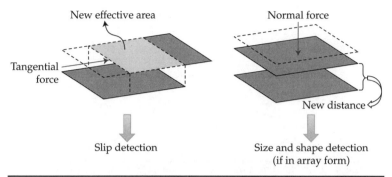

FIGURE 4.2 The effect of tangential and normal forces on the capacitor plates.

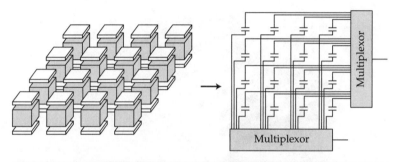

FIGURE 4.3 An array of capacitive sensors and how they are typically connected. (From Zhang, used by permission, 1999, Fig. 2.3.)

Placing multiplexors in the sensor array is considered a better approach. Figure 4.4 illustrates this design concept. Here, we have two sets of conductive strips. One of them serves as the static plates (*SP*) and the other acts as the moving plates (*MP*). At the intersection of each *SP* and *MP*, a tactile cell is generated. These intersections are denoted by SP_i and MP_j where i and j are indexes of the two conductive strip sets. Compared to the previous method in Fig. 4.3, this method reduces the number of connections when a large number of tactile cells are involved, thereby improving reliability.[1]

One possible application of these sensor arrays is in detecting shape, size, and slip as a complete touch sensation. To achieve this, we can place miniature slip detectors in these arrays between electrode gaps.[2] Here, there are no intersecting areas between the electrodes, as shown in Fig. 4.5.

Capacitive sensors are increasingly being used in industry and clinical applications, both in discrete and array forms. Their small size, easy fabrication, stability, low temperature dielectric coefficient, and

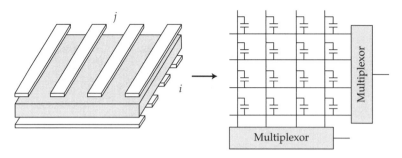

Figure 4.4 Embedded multiplexors in architecture of capacitive sensor array. (From Zhang, used by permission, 1999, Fig. 2.4.)

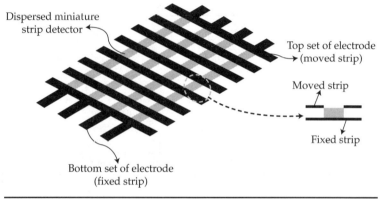

Figure 4.5 A capacitive array sensor with miniature slip detectors.

low noise make them more favorable than the other types of sensors.[2] As suitable alternatives for lost touch sensation, the capacitive sensors that have slip detectors are very promising when miniature pressure sensors are involved.

4.3 Inductive Sensors

Inductive sensors are mainly used to measure displacement, force, pressure, velocity, and acceleration. The operation of this type of sensor is based upon the induction of an electrical current in a magnetic field, depending on the magnitude of the mechanical stimuli.

The mechanical stimuli influence the total reluctance of the magnetic field; this is analogous to resistance in an electric circuit. Any change in reluctance leads to a change in induced current. Therefore, by measuring this change, the tactile property is quantified.

As a quick reminder, reluctance R (A·t/Wb) is the ratio of the magnetomotive force (mmf) to the total flux (Φ) as shown in Eq. (4.2):

$$R = \frac{\text{mmf}}{\Phi} \qquad (4.2)$$

The magnetomotive force is the total amount of current flowing through an N-turn coil, together with the total magnetic flux Φ (Wb) flowing through the cross section of a magnetic circuit. This can be expressed by Eq. (4.3):

$$\Phi = \int_A B \, dA \qquad (4.3)$$

where B is the magnetic flux density (T), and A is cross-section area (m²).

The total reluctance for a linear isotropic homogeneous magnetic material of length L and uniform cross-section area A can also be calculated from Eq. (4.4):

$$R = \frac{L}{\mu A} \qquad (4.4)$$

where μ is the permeability of the material (H/m). For a vacuum, air, or nonmagnetic material, μ is expressed as μ_0 and its value is $4\pi \times 10^{-7}$. For magnetic materials, μ is expressed as a product of μ_0 and relative permeability μ_R, which depends on the material. By examining Eq. (4.4), one can observe that the reluctances are small for magnetic materials with a larger μ_R. Similar to resistance in electrical circuits, reluctance is added when combined in series.

Depending on the mechanisms for altering this reluctance, inductive sensors are divided into different categories. One type is called a *variable air gap sensor* and is shown in Fig. 4.6. These types of sensors have ferromagnetic cores and armatures and, though very sensitive, are least affected by external magnetic fields.

Another type is the *inductive sensor with one or more coils and plunger-type armature*. Its structure is shown in Fig. 4.7. In most applications, however, variable air gap sensors appear to be more accurate and functional. Due to the large air path, the reluctance of the magnetic path in the coil and plunger-type armature sensor is high; therefore, its sensitivity is low.[2] Additionally, external magnetic fields can have a detrimental effect on these types of sensors. Also, in comparison to the variable air gap type, we would need more coil turns for obtaining a certain inductance.

Figure 4.8 shows the schematic of a possible inductive sensor with coil and a plunger-type armature. It is practical to put an elastic material inside of the coil and to support the T-shaped movable

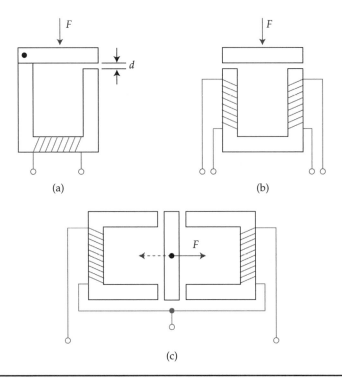

FIGURE 4.6 Variable air gap sensors with ferromagnetic cores and armatures.

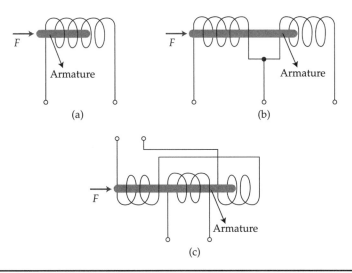

FIGURE 4.7 Structure of inductive sensors with one or more coils and a plunger-type armature.

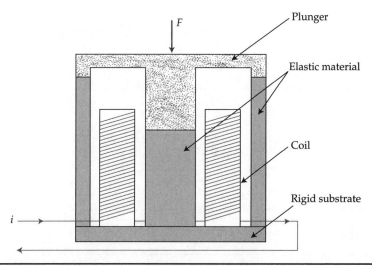

FIGURE 4.8 Schematic of an inductive sensor with coil and a plunger-type armature. (*Tactile Sensors for Robotics and Medicine*, J. G. Webster. Copyright © 1988 by John Wiley & Sons, Inc. Reproduced with permission of John Wiley & Sons, Inc.)

plunger. Changing the inductance of the plunger occurs when a load is applied on it and moves it into the coil.

Another type of inductive sensor is a *variable air gap sensor*, as shown in Fig. 4.9. The air gap changes as the applied force deflects the core itself; therefore, the reluctance varies. By comparison to other variable air gap sensors, a more uniform change of the air gap is obtained when using an E-shaped core. The elastic material can also help in determining the relationship between the applied force and the air gap. To do this, one can employ a linear elastic material placed between the moveable ferromagnetic armature and the coil with an E-shaped core.

Linear Variable Differential Transformer (LVDT)

One of the most popular reluctive displacement sensors is the *linear variable differential transformer* or *LVDT*. It is, in fact, an electromechanical sensor with mechanical displacement for its input and an AC carrier amplitude for its output. Here, the mechanical movement modulates the amplitude. With the distinct advantages of this type of inductive sensor, it has been used more widely than other sensors of this type in industry and research to measure pressure, displacement, and force. The main disadvantage of the LVDT is that it requires more complex signal processing instrumentation and circuits.[2]

A typical structure of an LVDT is shown in Fig. 4.10. On a cylindrical coil, we have three coils located at equal distances. The core of the assembly is a rod-shaped magnetic material whose purpose is

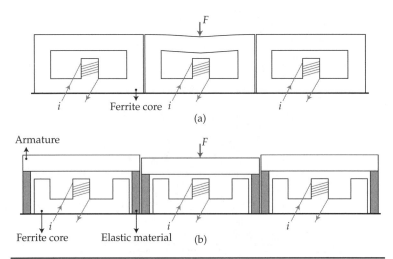

Figure 4.9 Array of variable air gap inductive sensors. (*Tactile Sensors for Robotics and Medicine*, J. G. Webster. Copyright © 1988 by John Wiley & Sons, Inc. Reproduced with permission of John Wiley & Sons, Inc.)

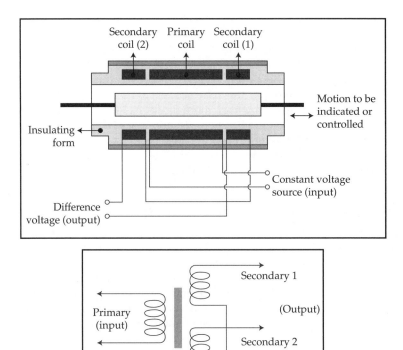

Figure 4.10 Typical structure of an LVDT. (See also color insert.)

to provide a path for the magnetic flux. Around a hollow nonmagnetic insulating coil form are the primary coil and the two secondary coils.[3]

Over a large range of core displacement, the LVDT can operate in a linear manner. At the ends, the LVDT saturates. When the core passes the center position, we observe a 180° phase change. This phase change shows the direction the core is moving. This moving pattern and phase change are shown in Fig. 4.11.

With the output waveforms that an LVDT produces, the direction of displacement cannot be distinguished if an ordinary rectifier-demodulator is used. Figure 4.12 shows this concept by illustrating two different input variables (x) producing the same output waveforms (V). To determine the direction of displacement, we need a phase-sensitive demodulator, whose function is shown in Fig. 4.13. In telecommunications, *modulation* is the process of varying a periodic waveform, such as a tone, in order to use that signal to convey a message. A device that performs modulation is known as a *modulator* and a device that performs the inverse operation of modulation is known as a *demodulator* (or sometimes as a detector or demod). Using this composition, the orientation of displacement can also be determined by an LVDT.

Some of the typical specifications of the LVDT include:

1. *Nominal linear range*: The maximal distance that the core can move in one direction and still be within the linear range

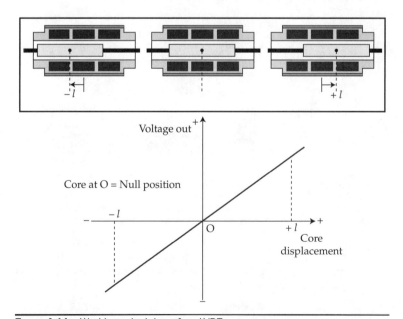

FIGURE 4.11 Working principles of an LVDT.

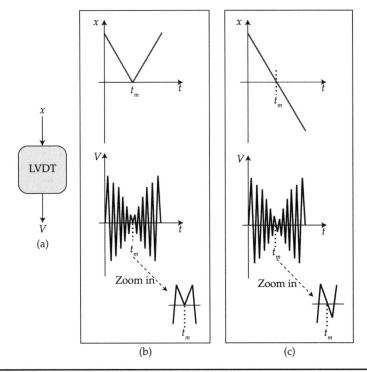

Figure 4.12 (a) LVDT sensor with input x (distance) and output V (voltage) (b) and (c) two same outputs per two different inputs.

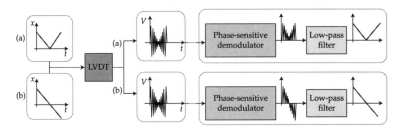

Figure 4.13 The role of a phase-sensitive demodulator in an LVDT sensor for determining the displacement orientation. (*Tactile Sensors for Robotics and Medicine*, J. G. Webster. Copyright © 1988 by John Wiley & Sons, Inc. Reproduced with permission of John Wiley & Sons, Inc.)

is called the *full-scale displacement*. Consequently, the linear operating range is twice the full-scale displacement. This is called the nominal linear range.[2]

2. *Linearity*: The maximal deviation of the output of an LVDT from a straight line is defined as the linearity of the LVDT.

This is provided that it is working within the nominal range. As a typical value, the nonlinearity of a standard LVDT is ±0.2% of its full output range.[4]

3. *Sensitivity*: Millivolts output per 0.001 inch of core displacement per volt of input ($mV_{out}/0.001$ inch/V_{in}) is normally considered to be the sensitivity of an LVDT.[2]

4. *Null voltage*: If the core is at the null position, the null voltage should be zero. However, in practice, the output voltage of commercial LVDTs does not become zero[2,3,5] at this position, but for many applications it can be considered negligible.

5. *Resolution and repeatability*: The smallest core position variation that can be detected at the output of the LVDT is called *resolution*. In other words, resolution is nearly infinite.[3,4,5] The electronic circuitry can limit the obtained resolution of the LVDT. Having mentioned that, it can measure very small changes of core position and the measurements can be repeated with great accuracy.

Based on the working principles and specifications of the LVDT, the authors have designed and fabricated a tactile sensor for detecting skin surface morphology, with applications in telemedicine systems. This will be introduced and described in the coming chapters.

4.4 Conductive Elastomers and Carbon Fibers

Conductive elastomer and carbon fiber sensors are both in the category of piezoresistive sensors. They have extremely nonlinear properties. The advantage of conductive elastomer sensors is that they can be used at high temperatures. They are also corrosion resistant. Other advantages include their low cost and their ability to be very small. The advantage of conductive carbon fibers is that they are strong, flexible, durable, and weigh much less than steel of similar strength. Additionally, they have good thermal stability and low hysteresis. The characteristics of these two types will be discussed in this section.

When we require materials with high elastic properties, one of the best choices is elastomers. These are polymers that can be greatly pulled or pushed. Their main feature is that they return to their original shape when the load is removed. Obviously, the loading process should be within their elastic limit. This is sometimes referred to as *large deformation* and their elastic properties as *hyper-elasticity*. Conductivity is added to them by impregnating the elastomers with metal powders or carbon black. They can be used as pressure sensors. As a result of the pressure that is applied on them, the electrical conductivity of the elastomers changes accordingly. A common type of these elastomers, often used today in sensors, is silicone-rubber.

The contact resistance between the conductive rubber and the metal plate affects the resistance of a conductive silicone-rubber sensor. The value of the resistance is at its maximum when there is no stress involved. Contrary to this, the resistance reduces when the system is compressed. This is due to an increase in the contact area.[2,3]

There are two ways that we can use these sensors. One is as discrete sensors and the other as arrays.[2] Figure 4.14 shows the isometric view of a half cylinder-shaped discrete sensor. When there is no load, the contact area is a line. If we apply pressure, then we see an increase in the rectangular area of contact. The graph shown in Fig. 4.15 shows the variation in the resistance of the sensor, assuming that its radius remains constant.

Figure 4.16 shows an hourglass-shaped sensor where the contact resistance changes between the two rubber surfaces. A single piece of rubber is used to make the sensor. Then, it is carved into an hourglass-shaped form. At zero pressure, the area of contact between the rubber surfaces is small. When we increase the pressure, the area increases. A wire wrap is used to make an electrical contact to the sensor. Here, the wire is soldered to copper tape and the tape is attached to the rubber with a conductive epoxy.[2]

As mentioned earlier, we can use conductive elastomers in array form. Figure 4.17 shows an array sensor made of carbon-loaded silicone rubber. By applying loads, the area of contact increases. This, in turn, leads to a reduction in the contact resistance.

Conductive rubber sensors have advantages of their own. They are appropriate for measuring high pressures since they can tolerate heavy overloads. Additionally, these types of sensors are strong in terms of enduring fatigue. Also, high temperatures and different chemicals cannot easily damage them. In addition, they can be built in very small dimensions, with minimal thickness and compliance. Their main disadvantage is that they exhibit high hysteresis.[2]

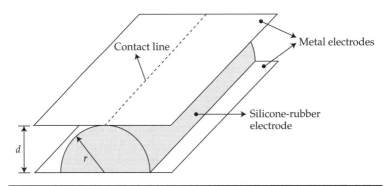

Figure 4.14 The front view of a half cylinder-shaped sensor based on conductive elastomer.

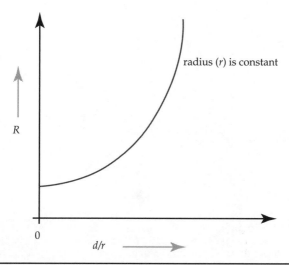

FIGURE 4.15 The variation in the resistance of a typical conductive elastomer. (*Tactile Sensors for Robotics and Medicine*, J. G. Webster. Copyright © 1988 by John Wiley & Sons, Inc. Reproduced with permission of John Wiley & Sons, Inc.)

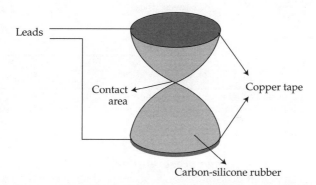

FIGURE 4.16 An hourglass-shaped sensor of carbon-silicone rubber. (*Tactile Sensors for Robotics and Medicine*, J. G. Webster. Copyright © 1988 by John Wiley & Sons, Inc. Reproduced with permission of John Wiley & Sons, Inc.)

The other type of piezoresistive sensors are carbon fiber sensors. Carbon fibers are made by carbonizing organic fibers and are available commercially as felt or rope. Fine cylindrical bundles of microfibrils are the main structural components of carbon fibers. In both forms of carbon fiber, rope or felt, the diameter of these microfibrils varies from 7 to 30 μm.[2] The physical structure of this carbon-fiber material enables its use as a sensor. This structure is shown in Fig. 4.18.

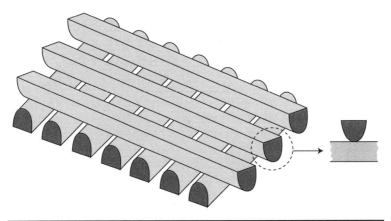

Figure 4.17 An array sensor made of carbon-loaded silicone rubber. (*Tactile Sensors for Robotics and Medicine*, J. G. Webster. Copyright © 1988 by John Wiley & Sons, Inc. Reproduced with permission of John Wiley & Sons, Inc.)

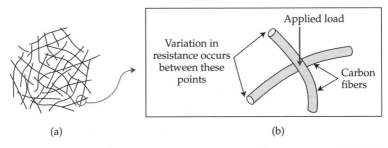

Figure 4.18 (a) Internal structure of carbon fiber containing microfibrils. (b) The variation in resistance occurs by the points of contact between the microfibrils in carbon fibers.

The resistance change occurs in three stages.[2] In Stage 1, the maximal change in resistance is observed. An increase in the number of contacts between the microfibrils in each bundle is responsible for this observation. Stage 2 deals with an increase in contact area between the microfibrils. The last stage, Stage 3, is because of an increase in contact area between the bundles themselves. When there is no load, the resistance of each bundle will depend on two factors. These are the number of microfibrils and the resistance of each microfibril.

Carbon fibers have a high tensile strength and stiffness, but at the same time are very flexible and can be fitted to any shape.[6] They can even be shaped to conform to the curves of a fingertip. Compared to conductive elastomers, carbon fibers have a low hysteresis and a high thermal stability and are the only fibrous material that can be used above 500°C.[2] These sensors seem well suited to monitoring

Introduction to Tactile Sensing Technologies 63

contact over large areas of a robot or for sensing in very inhospitable environments. The disadvantage of these sensors is that they generate noise and have poor shear strength.[7]

4.5 Optical Sensors

Optical sensors deal with light emitters and receivers. The sensors detect the presence, absence, position, and characteristic of objects by the interruption they cause in the light path, from the emitter through the receiver. This interruption results in a change in the intensity of the received light and, by measuring these changes, certain characteristics of the object can be established.[8]

The operating principles of optical sensors fall into two classes; these are *intrinsic*, where the optical phase, intensity, or polarization of transmitted light are modulated without interrupting the optical path; and *extrinsic*, where the physical stimulus interacts with the light external to the primary light path.[9] Both optical sensors can be used for touch and for force sensing.

Figures 4.19 and 4.20 show simplified schematics of how these sensors actually work. These sensors are called *opto-mechanical touch sensors*.[10] As is evident in the figures, the sensors are placed in such a way that the applied force on the object causes the object to come within the light path, therefore interrupting the light received by the receiver. This leads to different intensity of light emitted by the light emitting diode (LED) and light received by the photodiode (PIN). With this process, the magnitude of applied force, along with some characteristics of the surface of the object, can be determined.[8]

Some potential benefits of using optical sensors include: immunity to external electromagnetic interference, intrinsic safety, high resolution, low expense, light weight, and design simplicity.[11] However, since the optical sensors need a light emitter (source) and a receiver (detector), they cannot be made in small dimensions. In addition, they have a relatively slow response time of 10 to 100 ms,

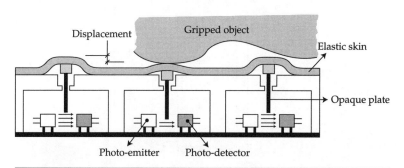

FIGURE 4.19 Optical sensors: An opto-mechanical array touch sensor.

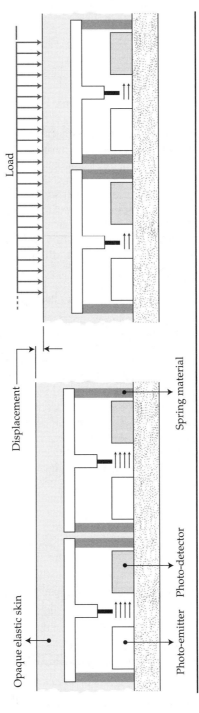

Figure 4.20 Optical sensors: Metal spring opto-mechanical touch sensor.

and they can only sense normal forces; therefore, they are not useful for shear force measurement. As a result of these restrictions, they are not very suitable for tactile sensory applications.

4.6 Thermal Sensors

The amount of heat flow out of a sensor into the object can be used as a suitable criterion for making a sensor. To measure applied pressure, a thermal sensor makes use of this property. For sensors with compliant surfaces, one can readily observe that the object-sensor contact area increases as the object is pressed more firmly against the sensor. The variation of sensor output is based on the fact that the heat flow out of the sensor is proportional to the contact area. A typical thermal sensor usually has three major components: heat source, a compliant surface, and temperature sensors.[2]

The above concept has been employed in the development of tactile sensor for robotics. However, for measuring pressure, it suffers from various disadvantages. One issue is related to the fact that thermal sensors are based on measuring a temperature difference between the object and the sensor. If this temperature gradient is relatively small, the sensor will show no significant output change. Second, for a given object shape and applied pressure, the sensor output depends on the object's thermal conductivity.[2] Therefore, objects of the same shape and size, but with different thermal conductivities, can display different outputs even though they are subjected to an equal magnitude of applied force. All of these problems raise serious concerns about the application of this type of sensor as a general pressure sensor, therefore limiting its application range.

4.7 Time of Flight Sensors

Another type of pressure sensor which can measure displacement caused by an applied pressure is called the *time of flight sensor*. These sensors measure the distance to an object by emitting a pulse of energy and measuring the time it takes for the reflected wave to return.[12] Figure 4.21 shows a simple diagram of the working principle of such a sensor. It has a wave transmitter and a receiver which are separated by a distance d. Compression springs can suitably keep them apart at this distance. An applied pressure compresses the springs and decreases the separation distance, making the travel time of the electromagnetic waves shorter. By measuring the change in the time of travel, the distance d is determined and, if the elastic constants of the springs are known, the applied force and pressure could be calculated.

An appropriate separation gap for a miniature sensor would typically be 1 mm. Assuming that the electromagnetic waves travel this distance with a velocity of 280 Mm/s, it would only take

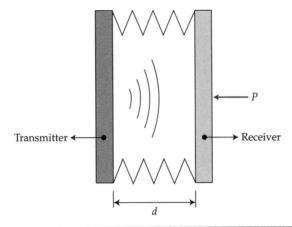

FIGURE 4.21 A schematic diagram of structure of a time of flight sensor. (*Tactile Sensors for Robotics and Medicine*, J. G. Webster. Copyright © 1988 by John Wiley & Sons, Inc. Reproduced with permission of John Wiley & Sons, Inc.)

3.57 picoseconds (ps) to travel the gap.[2] A change in distance resulting from a downward pressure requires a resolution of less than 3.57 ps, which can only be detected using precision instrumentation. For this reason, the practical application of this type of sensor is somewhat limited.

In order to tackle this problem, instead of using electromagnetic waves, we can utilize ultrasonic waves. Lower propagation velocities are the main characteristic of the ultrasonic waves. Rubber columns are normally used to separate the transmitter from the receiver. There are two reasons for this. One is that it reduces the interference in the air, and the other is that it is a good substitute for springs as seen in Fig. 4.22.[2] Let us say that the velocity of sound in silicone rubber at room temperature is 1.5 km/s. Therefore, for the rubber thickness of 1.1 mm, the travel time would be 0.733 μs. Here, a 1% change in distance can be measured using simple circuitry. This requires a resolution of 7 ns.

4.8 Binary Pressure Sensors

Switches are being used as active elements in binary pressure sensors. Due to their nature, they can only present responses limited to discrete values, such as *on* or *off* information.[2] Figure 4.23 shows two types of binary sensors. As shown, by applying pressure, we can deform the spring. This makes the rod move downwards. The mechanical properties of the spring determine the relationship between the pressure and displacement. Within a certain range, there is a linear relationship between the applied force and the displacement.

Introduction to Tactile Sensing Technologies   67

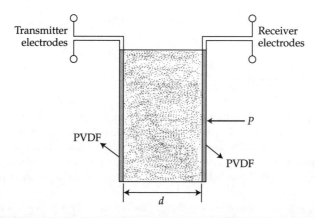

FIGURE 4.22 A schematic of time of flight sensor using ultrasonic wave and rubber columns. (*Tactile Sensors for Robotics and Medicine*, J. G. Webster. Copyright © 1988 by John Wiley & Sons, Inc. Reproduced with permission of John Wiley & Sons, Inc.)

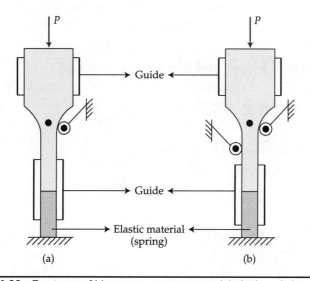

FIGURE 4.23 Two types of binary pressure sensors: (a) single switch, (b) double switch. (*Tactile Sensors for Robotics and Medicine*, J. G. Webster. Copyright © 1988 by John Wiley & Sons, Inc. Reproduced with permission of John Wiley & Sons, Inc.)

Note the place of the dot in Fig. 4.23 with respect to the switches. Fig. 4.23a shows that the sensor describes whether the dot is above or below the switch. As depicted in Fig. 4.23b, the sensor output can have three values; the dot can be between the switches, or it can be either above or below the upper or lower switch. Consequently, the resolution of the sensor can be determined by the number of switches.

4.9 Fluidic Coupling

A sensor that uses fluidic coupling is composed of four main parts: (1) a compressible bladder or other force-to-pressure sensor, (2) a remotely located pressure sensor, (3) a tube connecting the two, and (4) a fluid that fills the system.[2] A schematic of these compartments is shown in Fig. 4.24. By exerting force on the bladder, we increase the pressure of the fluid. The pressure sensor incorporated in the system can easily pick up the increase in pressure. Air, water, or oil could be used as the working fluid. The two main properties of fluids that play important roles in dynamic measurements are viscosity and compressibility. However, for static measurements, most fluids would be suitable.

Leakage is a possible source of error associated with fluidic systems. Temperature changes could also be a main source for error, as they could affect the viscosity and other characteristics of the fluid and subsequently cause pressure changes within the fluid.[2]

4.10 The Hall Effect and Magnetoresistance

The *Hall effect* takes place when an electric current is passed through a conductive material while the material is subjected to a magnetic

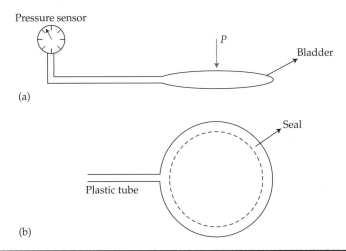

FIGURE 4.24 A fluid-filled bladder transmits pressure from a sealed bladder to an external pressure sensor: (a) side view; (b) top view. (*Tactile Sensors for Robotics and Medicine*, J. G. Webster. Copyright © 1988 by John Wiley & Sons, Inc. Reproduced with permission of John Wiley & Sons, Inc.)

field, mutually perpendicular to one another. Figure 4.25 shows an arrangement used to demonstrate this effect. In this example, a rectangular slab of conductive material is placed in a uniform magnetic field. The output from the Hall effect is the *Hall voltage*, V_H, in which its direction is perpendicular to both the electric current and the magnetic field. This is shown more clearly in Fig. 4.26.

Magnetoresistance is the change in resistance of a current-carrying conductor when a magnetic field is applied. A two-terminal device having this characteristic is called a *magnetoresistor*. In most magnetic materials the resistance of nonmagnetic conductors generally increases when a magnetic field is applied; however, electrical resistance decreases with the increase of magnetic field strength when the magnetic field direction is perpendicular to the current flow through the magnetoresistor. This concept, and its difference with the Hall effect, is shown graphically in Fig. 4.27.

Two forms of magnetoresistive sensors are shown in Fig. 4.28a and 4.28b. The one shown in Fig.4.28a is a magnetoresistive sensor using magnetic dipole while the other in Fig. 4.28b uses current-carrying wires to provide the magnetic field.

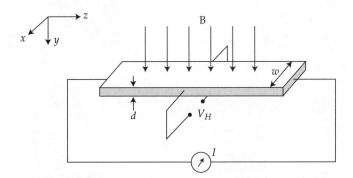

FIGURE 4.25 The Hall effect: Hall voltage (V_H) is measured across the left and right electrodes.

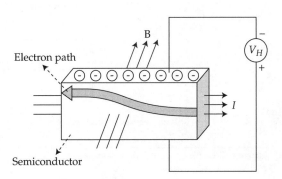

FIGURE 4.26 Schematic representation of the Hall effect.

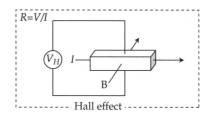

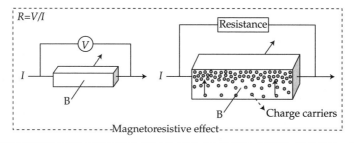

Figure 4.27 Magnetoresistance and its difference with the Hall effect.

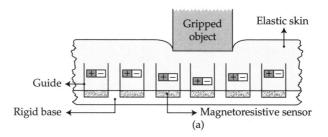

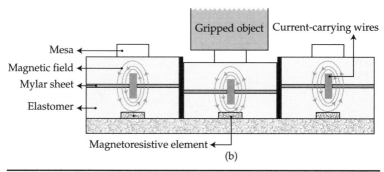

Figure 4.28 Magnetoresistive sensors: (a) using magnetic dipoles, (b) using current-carrying wires.

The Mylar sheet shown in this figure is a specific family of plastic sheet products made from the resin Polyethylene Terephthalate (PET). Both the Hall effect and magnetoresistance can be used to measure magnetic field intensity. For touch sensor designs, the magnetoresistive effect has advantages in terms of simplicity; it is, therefore, used more often.

So far, we have discussed the types of sensors that are more generally used in industrial and research applications. Their structures have been described and their working principles have been studied in simple terms. We have also gained some knowledge about the advantages and disadvantages of each type. With this knowledge, we can better judge their suitability for each application.

In addition to the described categories, there are two important types of tactile sensors which have broader and more profound applications: *piezoelectric* and *strain gauge sensors*. Due to their importance, these sensors will be discussed separately in more detail in the next two chapters. Additionally, some of the designed and fabricated sensors that are based on them and constructed and used by the authors, with specific applications in the biomedical industry, will also be introduced and reviewed.

References

1. P. Zhang, *Intelligent Gripper, with Multi-Sensor Finger Pad, for the Sarcos Dextrous Arm*, McGill University, Canada, 1999. (Dissertations & Theses, A&I Database, Publication No. AAT MQ50681.)
2. J. G. Webster, *Tactile Sensors for Robotics and Medicine*, John Wiley & Sons, Inc., 1988.
3. J. Fraden, *Handbook of Modern Sensors Physics, Designs, and Applications*, 3rd ed., Springer, 2004.
4. D. S. Nyce, *Linear Position Sensors Theory and Application*, John Wiley & Sons, Inc., 2004.
5. J. S. Wilson, *Sensor Technology Handbook*, Elsevier Inc., 2005.
6. J. B. Donnet, T. K. Want, S. Rebouillat, and J. C. M. Peng, *Carbon Fibers*, CRC Press, 1998.
7. B. D. Ratner, A. S. Hoffman, F. J. Schoen, and J. E. Lemons, *Biomaterials Science: An Introduction to Materials in Medicine*, Academic Press, 2004.
8. H. Singh, *Design, Finite Element and Experimental Analysis of Piezoelectric Tactile Sensors for Endoscopic Graspers*, M.A.Sc. Dissertation, Concordia University, Canada, 2004. (Dissertations & Theses, A&I Database, Publication No. AAT MQ94733.)
9. P. R. Nakka, *Design and Fabrication of a Micromachined Tactile Sensor for Endoscopic Graspers*, M.A.Sc. Dissertation, Concordia University, Canada, 2005. (Dissertations & Theses, A&I Database, Publication No. AAT MR04423.)
10. R. Andrew Russell, *Robot Tactile Sensing*, Prentice-Hall, December 1990.
11. B. Matic, *Sensor Technology for the Breast Examination Training Instrument*, M.S. Dissertation, West Virginia University, West Virginia, 2001. (Dissertations & Theses, A&I Database, Publication No. AAT 1407677.)
12. A. H. Slocum, *Precision Machine Design*, SME, 1992.

CHAPTER 5
Strain Gauge Sensors

5.1 Introduction

The strain gauge has been in use for many years and is the fundamental element for many types of sensors, including pressure sensors, load cells, torque sensors, position sensors, and others.

The majority of strain gauges are foil types, available in a wide variety of shapes and sizes to cater to a wide range of applications. They consist of a resistive foil pattern which is mounted on a backing material. They operate on the principle that, as the foil is subjected to stress, the resistance of the foil changes correspondingly.

5.2 Metal Strain Gauges

The function of these gauges is based upon the deformation of a piece of a metal by an applied force or stress. The induced strain thereafter changes the resistance; by measuring this change, the characteristics of the applied force can be determined.

The resistance of an object is related to the physical characteristics of that object as shown in Eq. (5.1):

$$R_0 = \frac{\rho l}{A} \rho \quad (5.1)$$

where R_0 is the resistance in ohms in unstrained state (Ω), ρ is the resistivity ($\Omega \cdot m$), l is the length (m), and A is the cross-sectional area (m^2) of the object. The dependence of resistance on strain is determined by the effects it has on the parameters in Eq. (5.1).

An incremental change in resistance R_0 can be expressed by Eq. (5.2):

$$dR_0 = \frac{\rho dl}{A} + \frac{l d\rho}{A} - \frac{\rho l dA}{A^2} = \frac{A(\rho dl + l d\rho) - \rho l dA}{A^2} \quad (5.2)$$

Chapter 5

The volume, V, is $V = Al$, and so dV would be equal to: $dV = Adl + ldA$. Strain, ε, is the change in length per unit length, and Poisson's ratio, v, is defined as lateral strain per longitudinal strain. For example, if the object is a bar it would have a square cross section and an applied tensile force along its length, as in Fig. 5.1. Assuming the original cross-sectional area is A, each side has a length of $A^{1/2}$. As a result of the tensile force F, each side contracts by an amount of $\varepsilon v A^{1/2}$, resulting in sides with a new of length $A^{1/2}(1-\varepsilon v)$. The new area and the new length of the bar can be expressed by Eqs. (5.3) and (5.4):

$$A_{new} = A(1-\varepsilon v)^2 \tag{5.3}$$

$$l_{new} = l(1+\varepsilon) \tag{5.4}$$

The change in volume can now be expressed by:

$$dV = A_{new}l_{new} - Al = l(1+\varepsilon)A(1-\varepsilon v)^2 - Al$$

By assuming that the strain, ε, is a small quantity and $\varepsilon v \ll v$, we would have:

$$dV = Al\varepsilon(1-2v) = Adl(1-2v) = Adl + ldA$$

From which we obtain Eq. (5.5):

$$-2vAdl = ldA \tag{5.5}$$

Substituting Eq. (5.5) into Eq. (5.2) leads to Eq. (5.6):

$$dR_0 = \frac{\rho dl(1+2v)}{A} + \frac{ld\rho}{A} \quad \rightarrow \quad \frac{dR_0}{R_0} = \frac{dl(1+2v)}{l} + \frac{d\rho}{\rho} \tag{5.6}$$

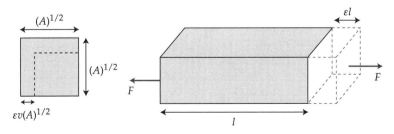

FIGURE 5.1 A bar with a square cross section under tensile pressure along its length.

Based upon this resistance, a parameter called the *gauge factor* is defined for metal strain gauges. The gauge factor, GF, is the sensitivity of the resistance to strain and can be determined by Eq. (5.7):

$$GF = \frac{dR_0 / R_0}{dl / l} \quad \text{or} \quad GF = 1 + 2v + \frac{d\rho / \rho}{dl / l} \tag{5.7}$$

By definition, the higher the gauge factor, the more suitable the material is for strain gauge applications. Higher values mean that the material is more sensitive to strain; even small values of strain (equivalent to small values of applied force) will change its resistivity. For most metals in their elastic region, Poisson's ratio is about 0.3 and that would usually lead to a gauge factor of approximately 2.

The material of construction for most metal strain gauges is an alloy of nickel and copper.[1] Constantan (copper-nickel alloy) and Isoelastic (nickel-iron alloy) are two of the most commonly used types.[2] Although Isoelastic gauges have higher gauge factors, they suffer from higher temperature sensitivities. Therefore, they can be used in measuring dynamic strains. In these cases, the temperature induced errors are not a major issue.

There are two main types of metal strain gauges, bonded and unbounded gauges.[1] The bonded gauge could be attached to the surface of an object. It experiences the same strain as the surface and is normally bonded to the measured surface. Glue or cement is usually used for the bonding process.

Metal foil, rather than wire, is normally employed as the strain-sensitive element. Using metal foils is advantageous because it is easier to manufacture. Additionally, it uses printed circuit-board construction technology. Figure 5.2 illustrates a schematic of a typical strain gauge.

The strain gauge has a finite length which makes measuring the strain at a single point impossible. Therefore, in practice, we are measuring the total strain averaged over the length of the active grid. The length of the grid region is also called the *gauge length*. This length is different from the length of the entire foil pattern. Obviously, if we make the gauge length smaller, measuring the strain at the desired point will be done with more accuracy. Despite this, larger gauges are frequently used instead. This is because making and handling smaller gauges are difficult tasks.

Figure 5.3 shows two examples of how a bonded strain gauge works. Figure 5.3a, shows a rectangular bar with a square cross section. The applied force is taken to be the tensile force in the x-direction. This causes tensile strain in the bar and we intend to measure this strain along the x-axis. The length of the bar increases under the applied force, inducing a strain in the bar; this creates a strain with the bonded strain gauge on the surface, yielding an increase in resistance within the gauge proportional to the force.

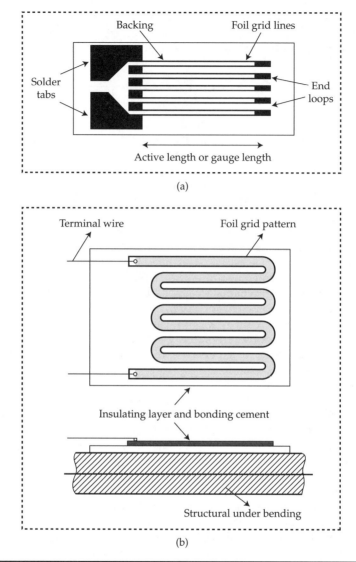

FIGURE 5.2 Schematic structure of strain-sensitive element and bonded gauges: a & b.

Figure 5.3*b*, shows a force applied on a cantilevered beam. In this case, the force in the *y*-direction causes the free end of the beam to deflect downwards, inducing a compressive strain on the bottom surface of the beam and a tensile strain on the top surface. The mounting location of the gauge is normally on either the top or bottom surface with its grid parallel to the *x*-axis. When the gauge is placed on top, increasing the applied force causes the resistance of

Strain Gauge Sensors 77

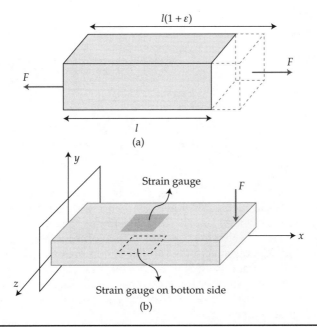

FIGURE 5.3 Strain gauge bonded on the surfaces on the objects under different types of loading: (a) tensile force leading to a tensile strain and (b) normal force causing deflection and inducing tensile and compressive strains on upper and lower surfaces, respectively.

the gauge to increase. On the bottom surface, however, the opposite occurs.

There are several benefits in using foil technology in bonded gauges. One advantage is that they could be fabricated in smaller sizes. The bonded gauge is only about 30 μm thick.[3] This includes the foil itself (3 to 8 μm) which is mounted on a plastic carrier (25 μm thick). Another advantage is their small cross-sectional area, which can be as small as 0.9 mm². They can also withstand larger strains, of up to 3%. Bonded gauges can also be used in harsh temperature environments.[3]

An unbonded gauge, on the other hand, consists of a thin wire of 0.002 cm or less wound between a fixed frame and a movable member.[4] Only wires can be used in this type of gauge. Therefore, several wires are usually employed to increase the resistance. This, in turn, leads to an increase in the magnitude of resistance change. An applied force changes the tension, causing a change in the resistance. The sensor is designed so that no expected force will cause the wires to become loose or relaxed. A schematic structure of this type of strain gauge is shown in Fig. 5.4.

The great advantage of the unbonded gauge is that it is very efficient in transmitting the applied force into a resistance change; no

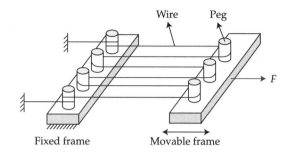

FIGURE 5.4
A schematic structure of an unbonded gauge.

force is borne by any member other than the gauge wires themselves.[3] One major disadvantage, however, is that it is difficult to produce a miniature unbonded gauge since the frames and pegs must be sufficiently large to bear the constant tension of the wires.[3]

Unbonded gauges were the most common pressure transducers prior to the availability of semiconductor strain gauges. This was due to their low hysteresis and creep characteristics. Since they have relatively large sizes, however, they have been replaced by smaller bonded strain gauges.

Based on the working concepts and principles of metal strain gauges, many force sensors have been designed and developed. They are commonly called *load cells* and are available in a wide variety of shapes, sizes, and load ranges from 1 N to more than 50 MN. However, very few are miniature. The dimensions vary from 6.3 to 12.3 mm in diameter and 2.8 to 6.3 mm in thickness.[3]

To understand how these force sensors are designed, let us view a simple example. Consider the sensor shown in Fig. 5.5. The beam is soldered to metal spacers on each end, deflecting like a diaphragm. The applied force causes the beam to deflect and produce tensile strain at the bottom surface of the center of the beam. The beam material is steel with elastic modulus $E = 200$ GPa. Assume that, due to the existing size and geometrical constraints, we have a supply of 0.30-mm-thick stock to use, we have a length of 12 mm, and we would like to measure up to 50 N. For this force, the strain should be 0.25%. Now both the beam width (b) and the beam deflection (Δ) must be calculated.

The bending moment at the center of this fixed beam could be calculated using Eq. (5.8):

$$M = \frac{FL}{8} \tag{5.8}$$

where F is the applied force and L is the length of the beam. The bending moment would thus be equal to Eq. (5.9):

$$M = \frac{\sigma I}{y} \tag{5.9}$$

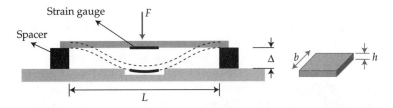

Figure 5.5 Schematic of a force sensor based on strain gauge. (*Tactile Sensors for Robotics and Medicine*, J. G. Webster. Copyright © 1988 by John Wiley & Sons, Inc. Reproduced with permission of John Wiley & Sons, Inc.)

In this equation, σ is the stress ($\sigma = E\varepsilon$), I is the moment of inertia about the z-axis, and y is the distance to the point from the center of the beam. This leads to:

$$\frac{FL}{8} = \frac{\sigma I}{y} = \frac{E\varepsilon I}{y}$$

For a rectangular cross section of our case, $y = h/2$ and $I = bh^3/12$, thus:

$$\frac{FL}{8} = \frac{E\varepsilon bh^2}{y}$$

Solving for b results in:

$$b = \frac{3FL}{4E\varepsilon h^2} = 10 \text{ mm}$$

So, the piece must be cut 10 mm wide.

Next, we must determine the maximal deflection, which is also the amount that the beam must be raised in the previous figure. For this beam, the deflection is calculated from Eq. (5.10):

$$\Delta = \frac{FL^3}{192EI} = \frac{FL^3}{192E(bh^3/12)} \tag{5.10}$$

$$\Delta = \frac{FL^3}{192E(bh^3/12)} = \frac{(50)(0.012)^3(12)}{192(200\times10^9)(0.01)(0.0003)^3} \rightarrow \Delta = 100 \text{ μm}$$

Thus, the beam shown in Fig. 5.5 should be mounted on 100 μm spacers. A force of 50 N would fully deflect the beam and cause it to contact the lower mounting surface. A recess in the lower mounting surface eliminates direct contact with the gauge. The strain at the center (also the maximal strain) would be 0.25%. Any further increase in force would not affect the sensor.

Figure 5.6 shows another sensor design which has overload protection. The metal arch deflects under an applied force and induces tensile strain on the underside surface where the strain gauge is mounted. For a certain applied force, the beam would fully deflect and lie flat. As in Fig. 5.5, the lower mounting surface is recessed to protect the gauge from direct contact. The sensor dimensions could be chosen so that this condition results in a strain of 0.25% so that the metal would not yield. However, for this type of force sensor design, scientists have reported that, in certain cases, a metal spring flexible enough to flatten against a stop did not produce enough strain.

The intrinsic characteristics of metal strain gauges are sources of errors associated with the measurements. Some of the important ones can be classified as follows:[3]

1. *Transverse sensitivity*: Strain would occur in both the longitudinal and lateral directions even if we have only longitudinal force. Therefore, the strain gauge will experience both strains. In practice, it would be more desirable if one could have a large gauge-resistance change for parallel strain and a small gauge-resistance change for transverse strain. Consequently, to eliminate this error, a low transverse sensitivity is required.

2. *Effects of ambient temperature*: In ideal cases, temperature should not have any effect on the resistance and gauge factor of a metal strain gauge. In reality, this does not happen and leads to errors in the measurements. It is advisable to select the gauge material so that it has a thermal expansion coefficient similar to that of the specimen. This decreases the effect of the thermal expansion difference among various components of the system.

3. *Self-heating*: A current must be passed through the gauge in order to measure its resistance. This causes heat generation

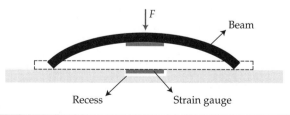

FIGURE 5.6 Schematic of force sensor based on strain gauge with overload protection design. (*Tactile Sensors for Robotics and Medicine*, J. G. Webster. Copyright © 1988 by John Wiley & Sons, Inc. Reproduced with permission of John Wiley & Sons, Inc.)

and the temperature problems discussed above. Additionally, we will see an undesired sensitivity to air flow around the gauge. To avoid these problems, we can decrease the gauge current, increase the thermal conductivity between the gauge and the specimen, and isolate the gauge from air currents.

4. *Lead wire resistance*: Even at a constant temperature, the resistance of the lead wires to the gauge terminals generates an effect which is known as *lead wire desensitization*. This decreases the effective gauge factor. Hence, we should keep the lead wires as short as possible.

5.3 Semiconductor Strain Gauges

Semiconductor strain gauges have high gauge factors compared to metal strain gauges and are, therefore, more sensitive to small strains. The gauge factor of a semiconductor strain gauge is between 50 and +200[5], in contrast to metal and foil strain gauges which were about 2. Although semiconductor strain gauges have a very high gauge factor, they are inherently temperature sensitive and nonlinear,[5] so compensation methods are necessary.

Currently, silicone is widely used in semiconductor strain gauges. Germanium, compounds gallium arsenide (GaAs), and gallium-antimony (GaSb) are other materials which are sometimes used in these gauges.[6] The advantage of using silicone in integrated circuits is that it makes the instruments more predictable and reproducible.

Silicone has an elastic limit greater than that of stainless steel and can tolerate cyclic loads. Additionally, it can be used at frequencies in the megahertz range and can withstand pressures greater than 350 MPa.[3]

Silicone has many useful and valuable characteristics. It resists corrosion, tolerates wide temperature ranges, and retains measurement accuracy of better than 0.1% even after use for many years.[3] Mass production provides uniformity from device to device. Thus, silicone is almost invariably used for semiconductor strain gauges. Most semiconductor strain gauges are now being manufactured by diffusing the chosen impurity into a thin surface layer of silicone wafer. This impurity has high resistivity or has an opposite conductivity.[3]

The equation for the gauge factor of semiconductor strain gauges is the same as that for metal strain gauges:

$$GF = \frac{dR_0/R_0}{dl/l} = 1 + 2v + \frac{d\rho/\rho}{dl/l}$$

where v is Poisson's ratio, ρ is resistivity, and l is length. Equivalently,

$$GF = 1 + 2v + \lambda_1 E$$

where λ_1 is the longitudinal piezoresistive coefficient (Pa^{-1}) and E is Young's modulus (Pa).[3] The gauge factor of semiconductor strain gauges predominantly depends upon the change due to the variation in resistivity with strain ($\lambda_1 E$). E and λ_1 depend on the material chosen and crystallographic orientation.

In contrast to metal strain gauges, the sensitivity of semiconductor strain gauges depends on the strain level. Therefore, the relation between resistance and strain is nonlinear, especially at high strain levels.[3] In general, the equation for sensitivity can be written as shown in Eq. (5.11):

$$\frac{dR_0}{R_0} = c_1\varepsilon + c_2\varepsilon^2 \qquad (5.11)$$

where c_1 and c_2 are constants, R_0 is the unstrained gauge resistance, and $\varepsilon = dl/l$ is the strain.

The advantages of silicone gauges include: high fatigue life and high strain sensitivity as compared to metal gauges; linearity of stress-strain curve; negligible hysteresis; less drift caused by moisture absorption; less noise due to imperfect installation of lead wires; and repeatable resistivity.[3]

Figure 5.7 shows a tactile sensor based on a semiconductor strain gauge designed and developed by the authors for medical applications. It is an integrated force-position tactile sensor for improving diagnostic and therapeutic endoscopic surgery. The designed assembly consists of two semiconductor microstrain gauge sensors, which are positioned at the back face of a prototype endoscopic grasper. The sensor can measure, with reasonable accuracy, the magnitude and the position of an applied load on the grasper. The intensity of the magnitude of the applied force to the endoscopic grasper can be visually seen on a light emitting diode (LED) device.[7]

For this application, silicone rubber was utilized to insulate the microstrain gauges; this allowed the assembly to operate in wet environments so common in most medical conditions without any difficulty. The authors have tested this by leaving the endoscopic grasper in a physiologic saline solution (0.9%) for 24 hours, after which time the performance of the device was not affected.[7]

Additionally, the designed system can be micromachined and, hence, integrated with endoscopic graspers of various dimensions. Furthermore, it is possible to miniaturize the complete electronics and place the sensor on the various endoscopic tools. While the designed device is of direct use in operating theaters by surgeons, this system has another potential application; it can be used in medical robotics telesurgery procedures, which are mainly based on teletaction operations.

Strain Gauge Sensors

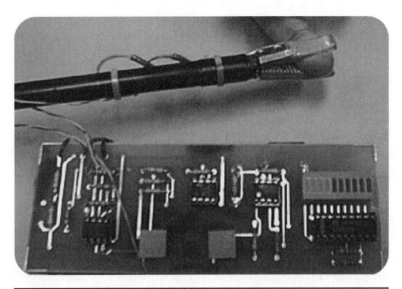

FIGURE 5.7 An endoscopic grasper and the associated electronic feedback system. The designed tactile sensor is mounted on the grasper. (Dargahi and Najarian, "An Endoscopic Force-Position Sensor Grasper with Minimum Sensors," *Canadian Journal of Electrical and Computer Engineering*, © 2003 IEEE.) (See also color insert.)

Currently, the entire operation during endoscopic surgery is viewed on monitors by the surgeons. Additional tactile information displayed on the monitors, such as stress or strain distributions in their crude forms, may be confusing to the surgeons; however, the visual perception of the intensity of the applied force magnitude displayed by an LED device (as demonstrated in this research work) will provide easier perception of the applied force.

Other advantages of the designed system are its simplicity and the use of a minimum number of strain gauges. By keeping the number of gauges down to an acceptable minimum, the number of wires connected to the gauges will also be reduced considerably as opposed to those used in array type of sensors. This, in turn, will result in a less bulky, and more practical, device. Therefore, this system can potentially be used in operating theaters without causing any unnecessary complications. At the same time, and based on the results obtained experimentally, the tactile sensor assembly is quite capable of estimating the location and magnitude of the applied force.

Figure 5.8 demonstrates a commercially available endoscopic grasper which has been equipped with an LED device close to the handle that shows the intensity of the applied force magnitude.[8]

In this chapter, we reviewed some of the most important types of strain gauge sensors. Currently, various researchers are focusing on

Chapter 5

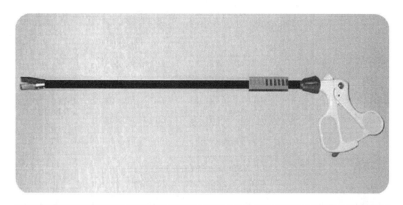

FIGURE 5.8 A commercially available endoscopic grasper, which has been integrated with the designed tactile sensor assembly. The miniaturized on-board electronics could be placed under the LED. (Reprinted from *Bio-Medical Materials and Engineering*, vol. 14, no. 2, Dargahi and Najarian, "An Integrated Force-Position Tactile Sensor for Improving Diagnostic and Therapeutic Endoscopic Surgery," pp. 151–166, Copyright 2004, with permission from IOS Press.)

improving the performance of the existing sensors. Based on these investigations, the future of using the improved versions of these sensors is quite promising.

References

1. Ch.Y. Ou, *Deep Excavation: Theory and Practice*, Taylor & Francis, 2006.
2. F. S. Tse, and I. E. Morse, *Measurement and Instrumentation in Engineering: Principles and Basic Laboratory Experiments*, CRC Press, 1989.
3. J. G. Webster, *Tactile Sensors for Robotics and Medicine*, John Wiley & Sons, Inc., 1988.
4. M. G. Joshi, *Comprehensive Transducers for Instrumentation*, Firewall Media, 2005.
5. C. R. Rao, and S. K. Guha, *Principles of Medical Electronics and Biomedical Instrumentation*, Orient Longman, 2001.
6. United States National Aeronautics and Space Administration, *NASA Technical Translation*, National Aeronautics and Space Administration, 1979.
7. J. Dargahi and S. Najarian, "An Endoscopic Force-Position Sensor Grasper with Minimum Sensors," *Canadian Journal of Electrical and Computer Engineering*, vol. 28, no. 3/4, 2004, pp. 155–161.
8. J. Dargahi, S. Najarian, "An Integrated Force-Position Tactile Sensor for Improving Diagnostic and Therapeutic Endoscopic Surgery," *Bio-Medical Materials and Engineering*, vol. 14, no. 2, pp. 2004, pp. 151–166.

CHAPTER 6
Piezoelectric Sensors

6.1 Piezoelectric Materials

Piezoelectric materials produce an electrical voltage when they are subjected to mechanical stress or deformation, and their dimensions change when they are subjected to an electric field. On the microscopic level, the structure of a material normally consists of uncharged randomly arranged dipoles. In a piezoelectric material, the dipoles are arranged along a single axis. When the material is subjected to a mechanical stimulus, such as stress or deformation, the dipoles shift from the axis, rearrange, and cause the charges to become unbalanced. This produces an overall electric dipole. The total charge of the material is the sum of all the unbalanced microscopic dipoles. This production of an electric charge by a material upon which a mechanical force is exerted is known as the *piezoelectric effect*.

In the late 1800s, Jacques and Pierre Curie were the first to discover the piezoelectric effect in crystals. Crystals have a very rigid structure, so large forces must be exerted on them to produce a charge.[1] Some very common crystals which exhibit this effect are quartz, tourmaline, and Rochelle salt.

6.2 Piezoelectric Ceramics

Under normal temperatures, a ceramic has a crystalline structure with no center of symmetry. At this state, the dipoles are randomly located in this structure in such a way that the overall dipole moment of the ceramic is zero as seen in Fig. 6.1a. By applying heat to the ceramic up to a certain temperature, its crystalline structure rearranges to a new structure with a definite center of symmetry. This temperature is called the *Curie temperature*; at this temperature, the dipole moment of the ceramic is still zero. By letting the ceramic cool down below the Curie temperature, the microscopic dipoles once again revert back to their random manner. By applying an electric field to the ceramic

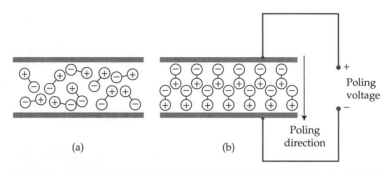

FIGURE 6.1 (a) Randomly directed dipoles in ceramic structure and (b) alignment of dipoles in the direction of applied electric field.

while simultaneously letting it cool below the Curie temperature, we see the alignment of the microscopic dipoles in the direction of the applied electric field, as seen in Fig. 6.1b. Now, we let the ceramic cool down to normal temperatures. At this stage, the electric field can be removed and the microscopic dipoles maintain their new alignment. This process, called *poling*, causes an overall dipole moment to appear in the ceramic.[2]

Under normal operating conditions, the microscopic dipoles of a poled ceramic cannot return to their random arrangement. The resulting ceramic is a permanent piezoelectric material. The variation of the dipole moment of the ceramic's structure leads to the generation of a voltage whenever the ceramic is deformed.

The advantage of a piezoelectric ceramic is that it can be made into almost any size or shape. This is not the case for a crystal. Therefore, piezoelectric ceramics are more often used in sensors. For the same amount of applied force, ceramic produces a greater electric charge. This is because ceramic is much less rigid than crystal. For instance, barium titanate ceramic (barium titanate is the principal ceramic piezoelectric) generates 140 pico-Coulombs per Newton (pC/N) compared to 2 pC/N for quartz.[*,2] Other advantages of a piezoelectric ceramic are: high durability, high tensile strength (approximately 80 GPa), and the choice for selecting the polarization direction.[2] In conclusion, for sensor designers, piezoelectric ceramic is a more favorable choice.

6.3 Directional Dependence of Piezoelectricity

Depending on the poling process, a ceramic material might have different piezoelectric characteristics in various directions. Let

---

*pC (pico-Coulomb) is a unit of electric charge.

us begin by defining a *coordinate system*. Constants that give an indication of a material's piezoelectric qualities are of the form g_{mn} where g is the directional piezoelectric constant. The two subscripts m and n give directional information about g. Here, m identifies the direction of the electric field, while n identifies the direction of the mechanical stress.

Consider the coordinate system shown in Fig. 6.2. A constant with the subscript 32 (i.e., g_{32}) gives the value of g described by the constant of a mechanical distortion applied in the "2" direction resulting in a voltage generated in the "3" direction. Therefore, unique voltage levels are being produced due to mechanical distortions in different directions.

The g constant is a measure of the generated electric field by the applied mechanical stress. The units of g are $(V/m)/(N/m^2)$ or $V \cdot m/N$ and the relation can be formulated as shown in Eq. (6.1):

$$E_m = g_{mn}\sigma_n \tag{6.1}$$

where σ_n is the applied mechanical stress (N/m^2 or Pa).

The electric field can also be calculated as the voltage per thickness of a piezoelectric material as shown in Eq. (6.2):

$$E = \frac{V}{t} \tag{6.2}$$

Therefore, combining Eqs. (6.1) and (6.2), a simple equation for the induced voltage is that shown in Eq. (6.3):

$$V_m = g_{mn}\sigma_n \tag{6.3}$$

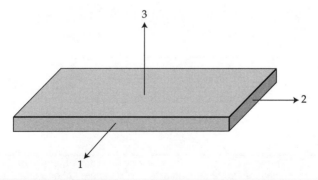

FIGURE 6.2 A coordinate system is needed for the description of mechanical deformation and induced voltages in piezoelectric sensors.

Consider the piezoelectric beam shown in Fig. 6.3a and note its poling direction. By applying a tensile force (resulting in tensile stress) in the "1" direction, a voltage (V) is generated in the "3" direction. With the same poling direction, if a compressive force (stress) is applied to the beam in the "1" direction, an opposite voltage ($-V$) is generated in the "3" direction, as seen in Fig. 6.3b. By using a beam with an opposite poling direction, the voltage polarity (V) would reverse. This is shown in Fig. 6.3c where a compressive stress has led to the same voltage as the tensile stress in Fig. 6.3a.

Piezoelectric sensors are mainly designed to work in the "31" mode, with operating principles similar to that illustrated in Fig. 6.3. However, in this mode, a piezoelectric sensor is not operationally

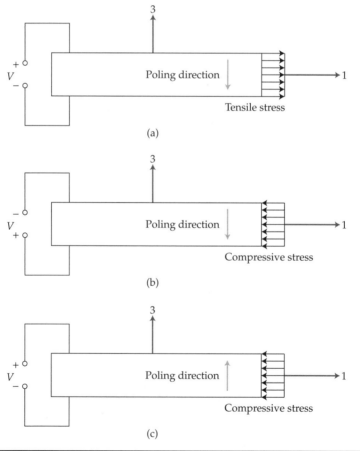

Figure 6.3 Type of stress and poling direction can affect the sign of the generated voltage: *(a)* beam under tensile pressure, *(b)* beam under compressive pressure, and *(c)* compressive pressure with the inverse poling direction.

very sensitive. To increase the sensitivity of the sensor in this mode, a special configuration of the piezoelectric sensor is used, known as *bimorph configuration*.[3] In this configuration, a higher output voltage is generated.

Two similar piezoelectric sensors are used in the bimorph configuration. They are poled for use in the "31" mode. The two piezoelectric strips are bonded together to form an element. This bonding is so that their poling directions are opposite to each other. If we apply force in the "3" direction, this element bends. As a result, the upper strip is in tension in the "1" direction and the lower one is in compression in the same direction (i.e., the "1" direction). A schematic of this configuration is shown in Fig. 6.4.

The mechanical strength of the piezoelectric sensor can be increased if needed. This is achieved by placing a thin metal strip between the two elements. The flexure element produced in this way is, in fact, a bimorph-configured piezoelectric sensor. Microphones, accelerometers, and strain gauges normally have a flexure element inside of them.[2]

In the case of applied forces in the "3" direction, the elements in the bimorph configuration are very effective. Therefore, mounting techniques such as those shown in Fig. 6.5a and 6.5b, are needed to cause the ceramic to be stressed in the "1" direction. If there is a need for a maximal deflection per force applied in the "3" direction, we

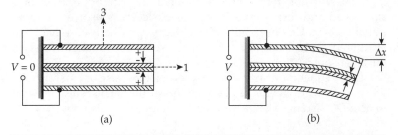

FIGURE 6.4 (a) The bimorph configuration consists of two identical piezoelectric sensors attached to a metal strip so that their poling directions oppose each other. (b) When deflected, the upper beam is in tension and the lower beam is in compression. (From *Tactile Sensors for Robotics and Medicine*, J. G. Webster. Copyright © 1988 by John Wiley & Sons, Inc. Reproduced with permission of John Wiley & Sons, Inc.)

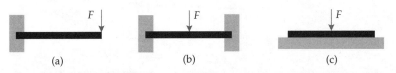

FIGURE 6.5 The bimorph beam: (a) one end supported, (b) two ends supported, and (c) mounted on fixed platform. (From *Tactile Sensors for Robotics and Medicine*, J. G. Webster. Copyright © 1988 by John Wiley & Sons, Inc. Reproduced with permission of John Wiley & Sons, Inc.)

can use the cantilever mounting shown in Fig. 6.5a. Another useful mounting technique called the *end-supported, center-driven technique*, is presented in Fig. 6.5b. Here, we fix both ends of the piezoelectric sensor. As shown in Fig. 6.5c, it is also possible to simply mount the sensor on a fixed platform. However, the output will be either zero or very small. This happens because element compression is in only the "3" direction with opposing polarities in the two layers of the bimorph.[2]

To obtain high sensitivity, a cantilever or end-supported mounting can be used. In this regard, the mounting process is of great importance. Should the element be exposed to unwanted forces from incorrect mounting, it can result in irreversible depolarization and fracturing of the element.

There are a number of applications for piezoelectric sensors. They can be used for many purposes, including to measure blood pressure, pulse rate, acceleration, and foot sole pressure. It should be noted that measuring static forces or pressures is not normally possible by a piezoelectric sensor. This is also true for a slowly changing pressure, such as barometric pressure.

Piezoelectric sensors also suffer from another disadvantage. They do not have a DC response.[4] This may have to do with the structure and nature of a typical piezoelectric material which causes it to act as a capacitor.[5] Capacitors can hold a charge, block direct current, and pass alternating current. Additionally, as with a capacitor, piezoelectric material will not hold a charge indefinitely. The application of a static load and displacements to a piezoelectric sensor leads to an initial voltage. However, this voltage decays gradually. This has been attributed to the existence of a leakage resistance of about 10 G in these sensors.[2]

The frequency response of a piezoelectric sensor is shown in Fig. 6.6. The sensor behaves like a high-pass filter.[6] This is also referred to as a *DC blocking filter*. A resonance frequency is also observed for

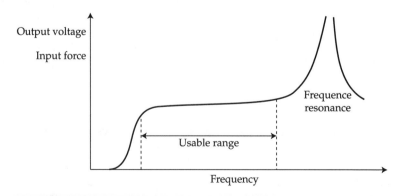

FIGURE 6.6 Frequency response of a piezoelectric ceramic. (From Cobbold, 1974. Reproduced with permission of John Wiley & Sons, Inc.)

this material, based on the frequency response output. According to this figure, the response is flat over a wide range of frequencies. One can use a piezoelectric sensor quite effectively in this flat range.

6.4 Polyvinylidene Fluoride

Polyvinylidene fluoride (polyvinylidene difluoride or PVDF or PVF_2) is a popular base resin used in the development of piezoelectric polymers. PVF_2 films have the same appearance as a thin sheet of plastic and are mechanically flexible and durable.[7] Their flexibility allows them to be manufactured into sheets as thin as 5 μm, or complex shapes depending on the application.[2]

PVF_2 is made up of dipoles that have been electrically aligned. This is done by poling in the same way as other piezoelectric materials. Therefore, the electrical response of PVF_2 is similar to that of other piezoelectric materials. Here, a voltage is generated across the surface of the film as soon as it is exposed to a compressive or tensile force. By unloading, we observe the generation of a voltage of reverse polarity in the material. Since the charge generated across the surface of the film decays through its internal resistance, polyvinylidene fluoride cannot be used for static measurements.

PVDF, like other piezoelectric materials, also exhibits a *pyroelectric effect*. This means that they are highly sensitive to temperature and an electrical output is generated when there is a change in temperature. Although this makes them useful for thermal measurements, this effect may be undesirable in some other applications. If we can control and isolate this pyroelectric effect from its piezoelectric effect, then the pyroelectric effect can be operationally useful and applicable. The pyroelectric voltage output can be calculated from Eq. (6.6)[2]

$$V = \frac{Ct\,\Delta T}{\varepsilon} \qquad (6.6)$$

where C is the pyroelectric coefficient ($C/m^2 \cdot K$), t is the film thickness (m), T is the absolute temperature (K), and ε is the dielectric constant of the film (F/m).

6.5 Piezoelectric Sensors in Biomedical Applications

Based on the characteristics and principles described, the authors have designed and developed a number of piezoelectric sensors for biomedical applications. The research activities have mainly focused on embedding artificial tactile sensing in surgical tools based on the piezoelectric effect. Some of these sensors are described in the following sections.

A Piezoelectric Tactile Sensor for Use in Endoscopic Surgery

The structural components of this sensor are shown in Fig. 6.7. It is a PVDF-based piezoelectric tactile sensor which could be integrated with an endoscopic grasper. The sensor exhibits high force sensitivity and linearity. Both computer modeling and experimental tests were performed to examine the function of this senor in different types of applied loading. A reasonable correlation was observed between the experimental and modeling results, with a difference between four to nine percent. The authors believe that a further study of the changes that occur in the multiple layers of the tactile sensor unit will lead to a better structural design in this type of sensor and yield more accurate output results.[8]

A Multifunctional PVDF-Based Tactile Sensor for Minimally Invasive Surgery

This is another tactile sensor system using PVDF. The working principle of this sensor allows for its use in combination with almost any end effectors; however, it is primarily designed to be integrated with Minimally Invasive Surgery (MIS) tools. In addition, the structural and transduction materials are selected to be compatible with Micro-Electro-Mechanical Systems (MEMS) technology, so that miniaturization could be possible. The corrugated shape of the sensor assures safe tissue grasping and compatibility with traditional tooth-like end effectors of MIS tools, shown in Fig. 6.8. A unit of this sensor is comprised of a base, a flexible beam, and three PVDF sensing elements as shown in Fig. 6.9. Two PVDF sensing elements sandwiched at end supports are used to measure the magnitude

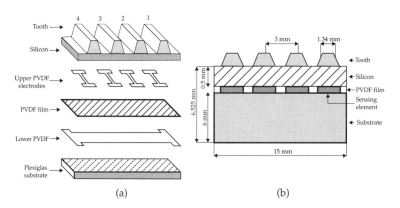

FIGURE 6.7 The designed piezoelectric tactile sensor for use in endoscopic surgery: (a) the structure of sensor and (b) the overall lateral view of sensor. (From Dargahi and Najarian 2004, courtesy of Emerald Group Publishing.)

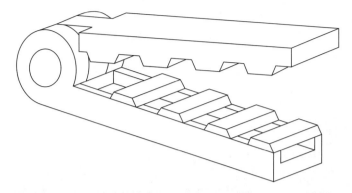

Figure 6.8 The MIS grasper equipped with an array of the proposed sensor. (From Sokhanvar, "A Multifunctional PVDF-Based Tactile Sensor for Minimally Invasive Surgery," 2007, IOP Publishing Ltd.)

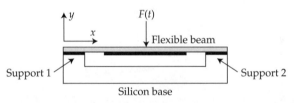

Figure 6.9 Cross section of the sensor unit (one tooth). (From Sokhanvar, "A Multifunctional PVDF-Based Tactile Sensor for Minimally Invasive Surgery," 2007, IOP Publishing Ltd.)

and position of the applied load. A third PVDF sensing element is attached to the beam and is used to measure the softness of the contact object. The tests on this prototype, which included analytical and numerical simulations, all proved that this sensor was capable of haptic sensing.[9]

To find the softness of an object, both the amplitude of the applied load and the resultant deflection are required simultaneously. A sensing mechanism utilizing PVDF films is attached at both end supports to measure the applied load, and to a flexible beam to measure the bending stresses. Knowing the applied load from the PVDF films at the supports, and the amount of deflection or the equivalent developed bending stresses from the middle of the PVDF film, it is feasible to characterize the softness of an object. The concept of the sensor operation is shown in Fig. 6.10.

As can be seen in Fig. 6.10a when a rigid object is used as target, there will be no resultant deformation in the beam. Although the PVDF films at the supports respond to the applied force, no output from the middle PVDF film is expected.

However, a soft object under the load F will bend the beam. As shown in Fig. 6.10b and 6.10c, larger beam deflection occurs for softer

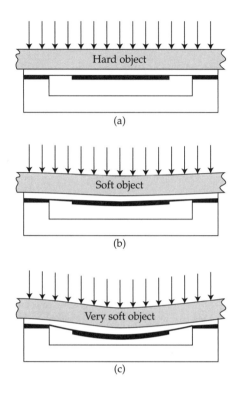

FIGURE 6.10
The basic idea of the proposed softness sensing technique: (a) hard object, (b) soft object, and (c) very soft object. (From Sokhanvar, "A Multifunctional PVDF-Based Tactile Sensor for Minimally Invasive Surgery," 2007, IOP Publishing Ltd.)

contact objects. Therefore, the PVDF film, which is firmly attached to the beam, elongates and develops a charge. This output charge is proportional to the bending stress, which itself is dependent on the extent of bending. The maximum amount of bending can be controlled in the design stage by selecting the material of the flexible beam as well as its thickness for a maximum given load. A beam with higher Young's modulus or thickness shows less deflection, but it can sustain higher applied load. Conversely, a soft beam deflects easily and produces a higher output charge, but it may not be a good choice for high loads.[9]

A Piezoelectric Tactile Sensory System with Graphical Display of Tactile Sensing Data

The goal of designing this system was to develop a commercial version of the endoscopic grasper equipped with a micromachined tactile sensor and with an associated signal processing and display system. For use as sensing elements, two PVDF piezoelectric films were incorporated into the sensing assembly. The combination of the resulting output voltages determined the object's softness. A feedback system was designed and incorporated into the sensor assembly which transmitted tactile signals from the grasper to the processing system. Following this step, we developed a signal processing and display system. By doing so, we managed to obtain the graphical representation of the tactile data on a computer monitor. The resulting data included the tissue softness and stress distribution on the tissue/grasper interface. Other important attributes, including the detection of an artery and lumps embedded within a complex tissue, can also be easily shown by color coding. Finally, the testing of the grasper device with its feedback and tactile imaging systems was performed. Various objects with known tactile properties were grasped with the device and the experimental outputs were compared with known values. An average discrepancy of about 10% was obtained when comparing the two sets of data. This system would be very effective in detecting a tumor embedded in biological tissue.

The sensor unit consists of a rigid cylinder surrounded by a compliant cylinder as seen in Fig. 6.11. The rigid and compliant cylinders were fabricated from Plexiglas and foam, respectively. The diameter of the rigid cylinder is 5 mm, while the diameter of the compliant cylinder is 1.5 mm. These two cylinders were glued on the top surface of one of the two Plexiglas bases. A 110 μm metalized uni-axially poled PVDF film from the Good Fellow Company, USA was put between these two Plexiglas bases, beneath both the rigid and the compliant cylinders.[10] Another PVDF film (the upper one) was put right under the rigid cylinder. When an object is in contact with the sensors incorporated into the system, a load is applied on both compliant and rigid cylinders. The softer the contact object, the more the transfer of load from the rigid cylinder to the compliant cylinder.

Figure 6.12a and 6.12b show the components of a designed sensor and the complete three-dimensional model of the whole end effector with eight tactile sensors.[11] Figure 6.13 shows a commercial endoscopic

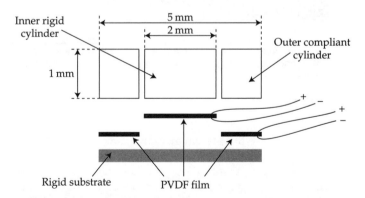

FIGURE 6.11 Cross-sectional exploded view of the sensor unit. (Dargahi and Najarian, "Graphical Display of Tactile Sensing Data with Application in Minimally Invasive Surgery," *Canadian Journal of Electrical and Computer Engineering*, © 2007 IEEE.)

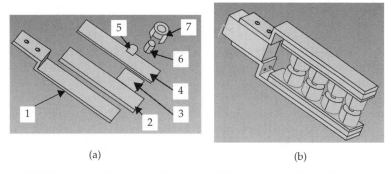

FIGURE 6.12 (a) Different parts of the designed sensor: 1-effector back-up; 2-base no. 1; 3-bottom PVDF; 4-base no. 2; 5-upper PVDF; 6-rigid cylinder and; 7-compliant cylinder. (b) Complete 3D model of the endoscopic grasper. ("A Novel Method in Measuring the Stiffness of Sensed Objects with Applications for Biomedical Robotic Systems," Najarian, et al. Copyright © 2006 by John Wiley & Sons, Inc. Reproduced with permission of John Wiley & Sons, Inc.)

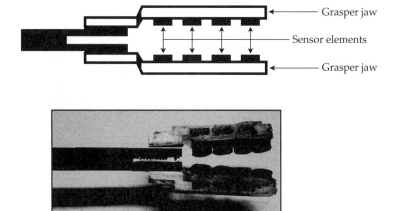

FIGURE 6.13 A commercial endoscopic grasper with four tactile sensors integrated with each jaw of the grasper. (Dargahi and Najarian, "Graphical Display of Tactile Sensing Data with Application in Minimally Invasive Surgery," *Canadian Journal of Electrical and Computer Engineering*, © 2007 IEEE.) (See also color insert.)

grasper equipped with this system including a micromachined sensor integrated on the grasper jaws.

The softness of the grasped tissue in this system is represented by color coding. As depicted in Fig. 6.14, a scale on the right-hand side of the graph which appears on the monitor shows the color coding numerically. In this case, the two right-hand sensors of the grasper are engaged. The softness display shows the sensed object in gray scale. Here, both the upper and lower sensors show the same softness. This

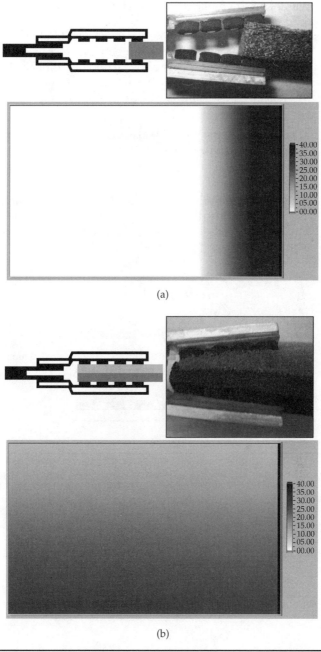

FIGURE 6.14 Measuring the softness on an object: (a) grasper is touching a uniform object; (b) upper and lower jaws are touching two different objects. (Dargahi and Najarian, "Graphical Display of Tactile Sensing Data with Application in Minimally Invasive Surgery," *Canadian Journal of Electrical and Computer Engineering*, © 2007 IEEE.)

means that the grasped object has the same softness throughout its thickness.

As previously mentioned, this system is capable of acting as a lump localizer. In order to differentiate between soft material and a hard lump embedded within the soft material, a threshold value technique is employed in this system. Any part of the object which is softer than this threshold value can be considered a lump that can be verified in two ways:

1. The position of the lump could be obtained by comparing the output of the sensors. The output of the sensor close to the lump is higher than other sensors. This is shown graphically in Fig. 6.15.
2. The position of the lump along the thickness of the embedded material could also be obtained by comparing the signals from the sensors positioned opposite from each other. Figure 6.16 explains this concept.

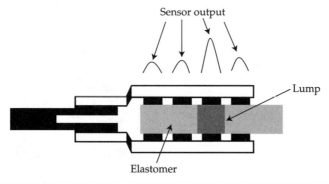

Figure 6.15 Localizing a lump along the grasper. (Dargahi and Najarian, "Graphical Display of Tactile Sensing Data with Application in Minimally Invasive Surgery," *Canadian Journal of Electrical and Computer Engineering*, © 2007 IEEE.)

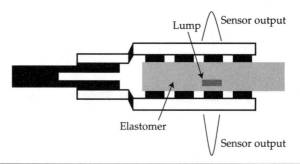

Figure 6.16 Localizing a lump across the grasper. (Dargahi and Najarian, "Graphical Display of Tactile Sensing Data with Application in Minimally Invasive Surgery," *Canadian Journal of Electrical and Computer Engineering*, © 2007 IEEE.)

In either case, an output graph is generated and will appear on the monitor. Figure 6.17 shows the graph of a lump embedded in the material, similar to case (2).

A Hybrid Piezoelectric-Capacitive Tactile Sensor

The proposed tactile sensor consists of a silicon substrate and a PVDF film sandwiched between the silicon substrate and Plexiglas membrane. The silicon substrate (as the base) has four cavities, each having a depth of 20 mm. A patterned PVDF film with a thickness of 25 mm is placed above the silicon substrate. The top Plexiglas was then structured to have a tooth-shaped profile, with a tooth thickness of 300 mm. Each tooth on the Plexiglas substrate was separated by 1.1 mm. The schematic and three-dimensional view of the sensor is shown in Fig. 6.18*a* and 6.18*b*.

Fabrication of this sensor requires a specific and delicate procedure. This procedure is briefly explained below.

The tactile sensor requires four masking steps regardless of its geometric parameters. Each masking step should take the required size and tolerance into consideration. The mask templates required for processing the silicon wafers, aluminum deposition, patterning PVDF, and patterning Plexiglas are shown in Fig. 6.19. The three main parts of the sensor, silicon structure, PVDF film and tooth-like Plexiglas are then fabricated and processed in the following manner:

1. *Silicon Structure*: The silicon substrate has cavities etched using anisotropic etching of silicon in tetramethylammonium hydroxide (TMAH). Initially, the single crystal silicon wafer was cleaned using an oxidation process which produced a uniform 5000 A° layer of silicon dioxide (SiO_2) on the surface of the wafer. This oxide layer serves as the mask for the anisotropic etching of silicon. The required mask was made in order to etch four cavities on the silicon wafer using photolithographic techniques. Furthermore, alignment marks were added to the masks for later use in aligning the PVDF on top of the silicon wafer. Then photoresist was spun on the wafer and was, subsequently, soft baked at 90°C for 20 minutes.

Using TMAH, the etching was carried out at 90°C resulting in the formation of four cavities with alignment marks on the surface of the wafer. The depth of the etching was found to be 20 μm. Aluminum electrodes were deposited on the whole of the etched cavities in the silicon substrate; they were also deposited on a few other parts on the surface of the substrate, which are used for making connections with the bottom electrodes of the PVDF film. The aluminum was deposited using a vapor deposition technique. This involved placing a silicon wafer in a closed vacuum chamber, heating an aluminum filament coil to its boiling temperature and depositing aluminum on the surface of the wafer.

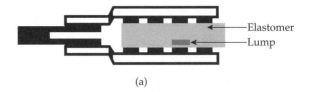
(a)

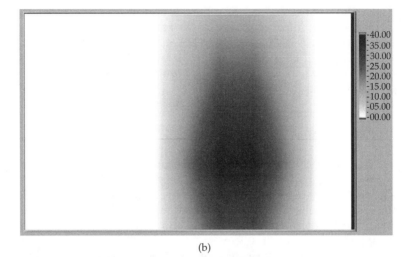
(b)

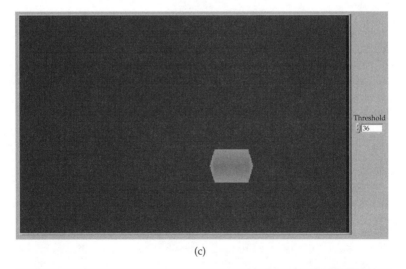
(c)

FIGURE 6.17 Detection of a lump within a soft tissue in color and gray scale modes: (a) a lump embedded within an elastomer, (b) lump detection, gray scale mode, and (c) lump detection, using threshold approach, color scale mode. (Dargahi and Najarian, "Graphical Display of Tactile Sensing Data with Application in Minimally Invasive Surgery," *Canadian Journal of Electrical and Computer Engineering*, © 2007 IEEE.)

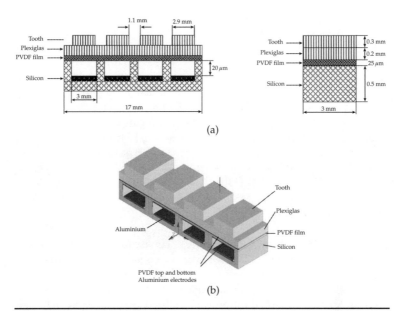

Figure 6.18 The structure of the hybrid sensor in which both piezoelectricity and capacitance variations have been taken into account: (a) 2D schematic of components and (b) 3D view of the tactile sensor. (From Dargahi 2006, courtesy of Emerald Group Publishing.)

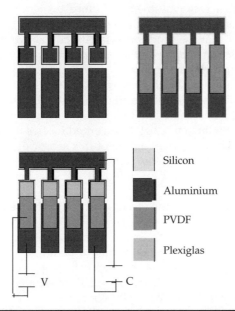

Figure 6.19 Different masks used in photolithography. (From Dargahi 2006, courtesy of Emerald Group Publishing.)

2. *PVDF Layer*: The second part is made from 25-μm poled PVDF film from Good Fellow Ltd, USA. The PVDF film was metalized with a thin coating of aluminum on both sides. The aluminum on the bottom side of the PVDF film was patterned with four strip electrodes. To do this, the film was carefully taped on the cleaned silicon so that the film was stretched uniformly over the wafer and the tape covered all the sides of the PVDF film. This ensured that the subsequent photoresist spinning did not cause the photoresist to creep under the PVDF film. The wafer was held by vacuum holding plates (vacuum chunk), and a few drops of photoresist were spun at 4000 rpm for thirty seconds; this resulted in an even layer of photoresist. Since the PVDF film loses some of its piezoelectricity if the temperature exceeds 70°C, the photoresist was semicured by soft baking at 60°C for one hour.[12]

The PVDF film was then carefully placed under the mask and exposed to UV light for a few seconds. The exposed PVDF film was then immersed in developer for forty seconds and then rinsed in deionization (DI) water. The wafer was air-dried and then cured in a hard bake oven at 60°C for another hour. The PVDF film was then inserted into commercial aluminum etchant mixture of which 73% was phosphoric acid, 11.5% acetic acid, 2.5% nitric acid, and 13% water until the patterned electrode strips were seen clearly. A rinse with acetone removed the photoresist layer, which was followed by a cleaning with DI water.[12]

3. *Tooth-like Plexiglas*: The third part was made of 0.5-mm-thick Plexiglas, which was patterned to form a 0.3-mm-thick tooth-like structure. Then, Plexiglas was glued to the top of the PVDF film in such a way that the tooth on it lay exactly above the patterned aluminum electrode on the bottom side of the PVDF. The PVDF film was glued onto the silicon substrate so that the patterned aluminum electrodes of the PVDF film, as well as the aluminum electrode deposited in the cavities of the silicon substrate, formed a parallel plate capacitor. The PVDF film, Plexiglas, and silicon were bonded using nonconducting epoxy glue and cured for five minutes at room temperature. The significant aspect of this sensor design is that the fabrication procedure was carefully planned to satisfy the primary objectives of the geometrical size and the pressure range of the sensor.

The experimental setup used for testing the sensor for both static and dynamic responses is shown in Fig. 6.20. It consists of a signal generator, which is the source of sinusoidal excitation, and an amplifier for the signal generator output before sending the signal to the shaker. The outputs of the microfabricated sensor, as well as another reference load cell, are passed through a charge amplifier before being recorded on the oscilloscope.

Due to the high sensitivity of PVDF film used in this sensor and the sensor's configuration, many of the possible errors and their

Piezoelectric Sensors

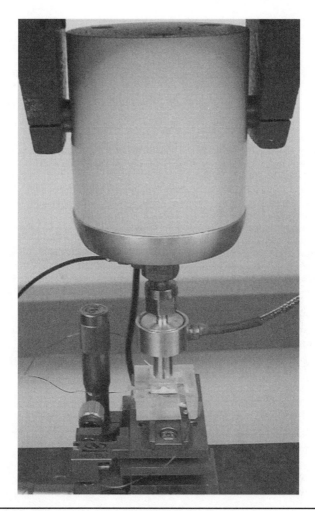

FIGURE 6.20 Experimental setup for testing the designed sensor. (From Dargahi 2006, courtesy of Emerald Group Publishing.)

sources which normally exist in other types of tactile sensors with the same applications, have been eliminated. The only main source of error in this system is restricted to the probable improper alignment of the sensor.

References

1. G. Gautschi, *Piezoelectric Sensorics: Force, Strain, Pressure, Acceleration and Acoustic Emission Sensors, Materials and Amplifiers*, Springer, 2002.
2. J. G. Webster, *Tactile Sensors for Robotics and Medicine*, John Wiley & Sons, Inc., 1988.

3. D. J. Leo, *Engineering Analysis of Smart Material Systems: Analysis, Design, and Control*, John Wiley and Sons, Inc., 2007.
4. M. G. Joshi, *Comprehensive Transducers for Instrumentation*, Firewall Media, 2005.
5. H. L. Trietley, *Transducers in Mechanical and Electronic Design*, CRC Press, 1986.
6. R. S. C. Cobbold, *Transducers for Biomedical Measurements*, John Wiley & Sons, Inc., 1974.
7. J. G. Webster, *The Measurement, Instrumentation, and Sensors Handbook: Handbook*, CRC Press, 1999.
8. J. Dargahi and S. Najarian, "Theoretical and Experimental Analysis of a Piezoelectric Tactile Sensor for Use in Endoscopic Surgery," *Sensor Review*, vol. 24, no.1, 2004, pp. 74–83.
9. S. Sokhanvar, M. Packirisamy, and J. Dargahi, "A Multifunctional PVDF-Based Tactile Sensor for Minimally Invasive Surgery," *Smart Materials and Structures*, vol. 16, 2007, pp. 989–998.
10. J. Dargahi, S. Najarian, and R. Ramezanifard, "Graphical Display of Tactile Sensing Data with Application in Minimally Invasive Surgery," *Canadian Journal of Electrical and Computer Engineering*, vol. 32, no. 3, 2007, pp. 151–156.
11. S. Najarian, J. Dargahi, and X. Z. Zheng, "A Novel Method in Measuring the Stiffness of Sensed Objects with Applications for Biomedical Robotic Systems," *International Journal of Medical Robotics and Computer Assisted Surgery*, vol. 2, no. 1, 2006, pp. 84–90.
12. J. Dargahi, M. Kahrizi, N. Purushotham Rao, and S. Sokhanvar, "Design and Microfabrication of a Hybrid Piezoelectric-Capacitive Tactile Sensor," *Sensor Review*, vol. 26, no. 3, 2006, pp. 186–192.

CHAPTER 7
Application of Tactile Sensing in Surgery

7.1 Open Surgery and Minimally Invasive Surgery

During open surgical operations, also sometimes referred to as *laparotomy* if performed on abdominal organs,[1] surgeons rely heavily on sensations from their fingertips to guide manipulation and to perceive a wide variety of anatomical structures and pathologies. Distinguishing between different tissues and organs during surgery is mainly achieved by the surgeon's sense of touch; this sense helps to determine how much force to exert by the hands and how to avoid injuring nontarget tissues.

Traditional open surgical procedures usually require large incisions (15 to 30 cm) to be made on body. The benefits to the surgeons in this type of surgery are many. They have a large visual field with direct visual and tactile contact with the area of surgery; they have direct tactile feedback; and they have freedom of motion due to the nature of such large incisions. Nonetheless, this type of surgery includes significant blood loss and long recovery times for the patients.[2] In addition, in order to keep the incision open for the duration of the operation, the surgeon is required to use metal fixtures. These may exert huge forces on the patient's body and the incision area. A schematic of the site of incision and open surgery in the body is presented in Fig. 7.1.

To reduce these disadvantages, a new revolutionary surgical technique known as *Minimally Invasive Surgery* (*MIS*) is being employed. This surgical technique, also sometimes referred to as *endosurgery*,[3] *endoscopic surgery*,[4] *keyhole surgery*,[5] and *minimally access surgery* (*MAS*),[6] integrates engineering and medicine with the goal of minimizing trauma to the patient. Surgeons gain access to the target area through small ports or incisions (about 1 cm) on the body surface, as shown in

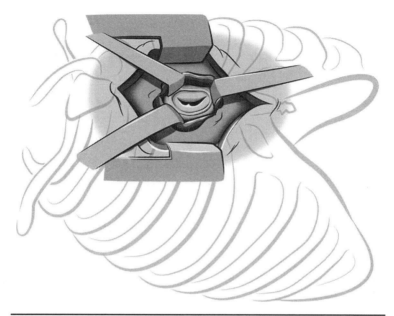

FIGURE 7.1 Schematic of an incision in open surgery. Metal fixtures are used to keep the incision open during the operation. (Photo courtesy of Intuitive Surgical, Inc., 2008.) (See also color insert.)

Fig. 7.2. The surgical instruments, which include a camera to provide vision to the surgeon, and a light source to produce enough light for the images to be captured by the camera, are inserted through these small incisions into the body. Since MIS involves less postoperative pain, minimal scattering, decreased rates of infection, shorter hospitalization time, and overall reduction in costs, this type of surgery is rapidly becoming popular for both surgeons and patients.

The areas of application of MIS are presently widespread and are growing. The diagram shown in Fig. 7.3 presents its different areas of application.

Colonoscopy is an examination for tumor identification in the colon. Examination of the chest cavity using an endoscope is referred to a thoracoscopy. Arthroscopy is the examination and repair of skeletal or joint disorders. Examination and unclogging of blood vessels is called angioscopy. Another technique in MIS is laparoscopy, which can be divided into four categories: cholecystectomy, inguinal hernia repair, appendectomy, and colectomy. Generally, laparoscopy means endoscopic surgery performed on abdominal organs. Gall-bladder removal is termed cholecystectomy. Inguinal hernia repair is the repair of a hernia in the groin. Removal of vermiform appendix is called appendectomy. Finally, colectomy is the removal of part or the entire colon.[3]

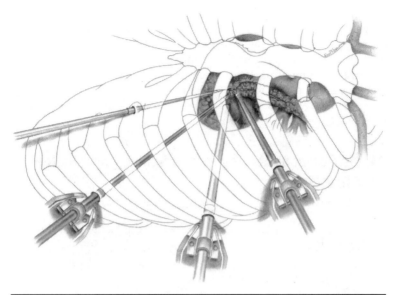

FIGURE 7.2 Surgical instruments are inserted into the patient's body through small incisions in MIS. (Photo courtesy of Intuitive Surgical, Inc., 2008.) (See also color insert.)

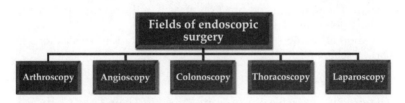

FIGURE 7.3 The areas of application of MIS.

While there are considerable advantages associated with MIS, there are also some drawbacks compared to traditional open surgeries. The main disadvantage is that the visual and tactile information available to the surgeon is greatly reduced in comparison to open surgery.[5–7] When MIS is adopted, the direct touch of the surgeon's hands with the patient's body is lost; therefore, any tactile feeling, to which the surgeon has become accustomed, almost vanishes. The surgeon's perception is limited to visual feedback from either a video camera or gross motion and force feedback through the handles of long instruments. In particular, the display in MIS cannot supply the surgeon with a sufficient feeling of depth, and is reliant solely upon the two-dimensional image on the screen.

In order to solve these problems, a considerable number of research activities are currently being undertaken. These efforts are mainly focused on using tactile sensory systems with MIS instruments in order to provide the surgeon with tactile feedback information from the surgical site. So far, a variety of tactile sensors have been developed which are suitable for mounting in a probe or surgical instrument. Their main aim is to simulate tactile stimulus directly on the surgeon's fingertips. To present the information from these sensors, one can send the data to display devices. In open surgeries, by using these remote palpation devices, some of the surgeon's perceptual and manipulative skills can be restored.[8]

7.2 Basic Components of a Tactile Sensing System for Use in MIS

A tactile sensing system for MIS comprises three basic parts:[9]

1. It includes a *tactile sensor* to extract the tactile data through contact.
2. It also includes a *tactile data* processing unit that processes the transduced data to obtain useable information.
3. Finally, it includes a *tactile display* unit that presents this information to the surgeon in a perceivable way.

The diagram in Fig. 7.4 shows the operational cycle of such a system. The structure and function of each of the above components is also briefly described.

Another tactile sensing system used in surgical applications is shown schematically in Fig. 7.5.

Tactile Sensor

The structure of tactile sensors is described in Chap. 4 and Chap. 5. A typical tactile sensor for use in MIS comprises four layers[9] as shown in Fig. 7.6:

1. A sensing layer
2. An electronics layer
3. A protective layer
4. A support layer

The sensing layer consists of either a single element or a one/two-dimensional array of elements. There are two types of tactile arrays. In the first type, a number of discrete sensor elements are connected together by wiring. In the second type, there is a continuous sheet of sensor material. Ideally, we want an array to be both sensitive to tangential forces and perpendicular forces. Some of the currently available transducer technologies are: piezoresistive, piezoelectric, optical, mechanical, and capacitive. Elastomer or silicon micromachining technology can also be used to produce certain transducers.[9]

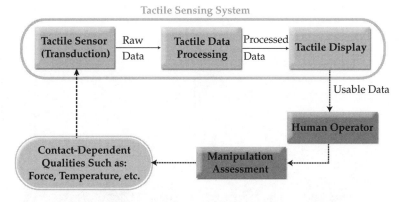

Figure 7.4 Working flowchart of a typical tactile sensing system for use in MIS. (Reprinted from *Mechatronics*, vol. 13, Eltaib and Hewit, "Tactile Sensing Technology for Minimal Access Surgery–A Review," pp. 1163–1177, Copyright © 2003, with permission from Elsevier.) (See also color insert.)

The role of the electronics layer is *signal conditioning*. This layer has all the essential hardware for the mentioned process. Signal conditioning involves the amplification and attenuation of a signal to prepare it in such a way that it meets the requirements for the next stage of further processing. Multiplexors, bridges, and amplifiers located in this layer are all part of functional circuitry. Using integrated circuit (IC) compatible technology, we can integrate this layer with the sensing layer in the same chip. Micromachining is a typical IC compatible technology that can be used.

A thin layer of elastic material is also used for protecting sensing and electronic circuitry elements. Although the presence of this layer is crucial for protection against sudden and unwanted external mechanical stresses, it may also complicate the analysis of the sensor output.

There is also another layer that acts as a base upon which the electronics and sensing layers are placed. This is called the *support layer*, and it may be either rigid or flexible. The method of attaching this layer to the gripper greatly affects the performance of the overall sensor system.

Tactile Data Processing

Data processing is an important aspect of any tactile sensing system because information can only be extracted by this means. A variety of algorithms for processing sensory data may be used. A common algorithm, solid mechanics, and finite element methods[9] can all be used for the construction of a forward model of a tactile sensor; they can also be used in order to obtain subsurface stress/strain data from the surface stress and material properties of the protective layer which covers the sensor. Additionally, an inverse model has been

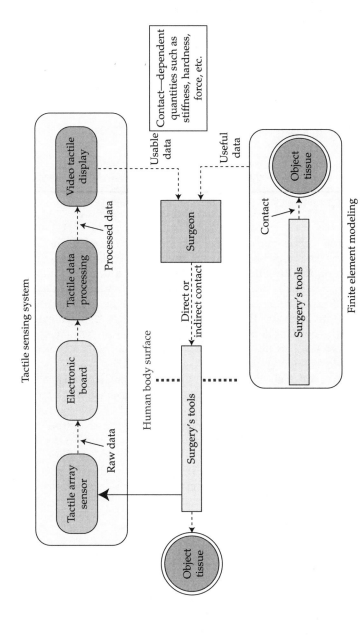

Figure 7.5 The components of a typical tactile sensing system and its role in surgery. (See also color insert.)

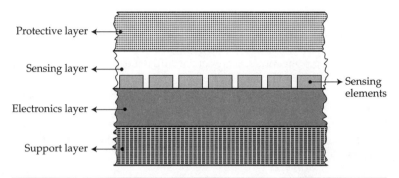

FIGURE 7.6 Structure of a typical tactile sensor for use in MIS. (Reprinted from *Mechatronics*, vol. 13, Eltaib and Hewit, "Tactile Sensing Technology for Minimal Access Surgery–A Review," pp. 1163–1177, Copyright © 2003, with permission from Elsevier.)

investigated in which the contact parameters are calculated from the subsurface strain. These strains are measured by the tactile sensor.

Tactile Display

The ultimate goal in using a tactile system in MIS is to sense tactile stimuli in the remote environment of the human body, and present them to the surgeon with the best possible fidelity, including high reliability and conformity. A display is used to present the extracted tactile data by the processing unit to the surgeon and includes information about texture, local shape, and local compliance.[9] Actuator size can limit the spatial resolution of a tactile display. Despite this, new possibilities have opened by the rapid developments in the field of microfabrication technologies. The main target is to meet the requirements of high-resolution graphic tactile display.

Design Considerations for Tactile Sensing Systems in MIS

A description of the working environment is of crucial importance for the design of a tactile sensor. This is due to the fact that the purpose of a tactile sensor is to measure or detect contact quantities between itself and the touched body. If we decide to make use of tactile arrays, then we should determine the type, the number, and the arrangement of the sensing elements forming the selected array. The mechanical properties of the environment for manipulation, including the various kinds of objects or biological tissues, are also very important in tactile exploration.[9] Some of these properties are weight, compliance, and viscosity. Some of the environmental parameters can vary including the locations and orientations of bodies together with their physical conditions such as wetness and temperature.

The working environment in MIS is a closed system containing soft tissue, living organs, and body fluids. In this environment, there are also various instruments used by the surgeon. The tactile sensor used for MIS will go inside the patient's body. Therefore, it must have certain special features normally encountered in biomedical engineering applications. The sensor must be miniature, reliable, biocompatible, waterproof, and disposable. Other related issues are cost and ease of assembly and disassembly. To address these issues, one can use elastomer-based tactile sensors. Even so, some inherent limitations still remain to be tackled. In order to meet all of the required specifications, the best choices so far are silicon-based tactile sensors.[9]

7.3 Remote Palpation Instruments for MIS

In order to provide the surgeon with palpation (see Section 8.1) in minimally invasive procedures, tactile sensors should be mounted on special surgical instruments that remotely measure tactile sensations from inside the patient's body and display this information to the surgeon. These remote palpation devices will gather tactile information necessary for diagnosis and localization.

Figure 7.7 illustrates a conceptual representation of a remote palpation instrument. Internal tissues can be probed through small entry incisions. This can be done by extending the length of the surgeon's finger. One solution is to use a remote palpation instrument. In ideal cases, the instrument would reproduce tactile sensations that replicate the sensation that the surgeon would actually feel if touching these inaccessible tissues directly.[5]

Tactile perception through palpation is composed of two distinct modalities: *tactile* or *cutaneous sensing* and *kinesthetic sensing*. The two

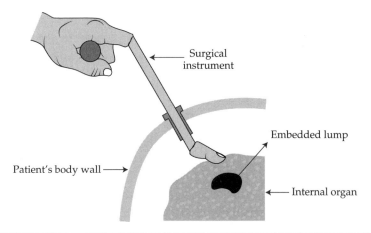

FIGURE 7.7 Working concept of remote palpation instrument for palpating internal organs. (From Peine, used by permission, 1999.)

Application of Tactile Sensing in Surgery   113

sensing modalities were introduced in Chap. 3. The combination of these sensations, often referred to as *haptics*, allows tactile exploration by integrating local shape or curvature information with overall finger location and contact force.

The structure of a typical surgical instrument capable of providing remote palpation for the surgeon is shown in Fig. 7.8a. The instrument consists of a number of sensors. These sensors are located at the tip of the instrument, and their purpose is to measure contact information from probed internal tissues. The characteristics that we are normally looking for include distributed pressure, shape, temperature, or vibration. These measured signals are being changed by a signal processing algorithm. Then, they are sent into drive commands for tactile display devices in the instrument handle. By so doing, sensations of the remote tissue are recreated directly on the surgeon's fingertip. Through a cable drive mechanism, the motions of the surgeon's finger are also transmitted to the tactile sensor on the instrument tip.[5] A prototype of such an instrument is shown in Fig. 7.8b.

The surgical instrument provides kinesthetic feedback by coupling the motions of the surgeon's finger with the motions of the

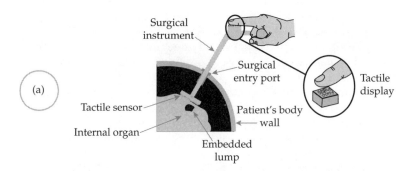

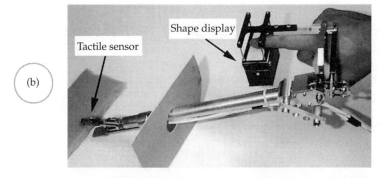

FIGURE 7.8 (a) Components of a surgical instrument for remote palpation and (b) a prototype of this instrument. (From Peine, used by permission, 1999.) (See also color insert.)

sensor. Internal organs are palpated by scanning the sensor probe across the tissue surface while feeling the induced tactile sensations.

The purpose for the design and fabrication of these devices is to provide clear and exact tactile feedback to the surgeon so that the surgeon has the same feeling, and is able to obtain the same amount of information, as if he or she had direct hand contact with internal tissues. Surgeons could then use this complete tactile perception while retaining the benefits of minimally invasive surgery.

However, the development of such tactile feedback systems is extremely challenging. Palpation is a difficult problem to study because it involves many complex components such as large-deformation contact mechanics, nonlinear biological materials, human tactile perception, and sensory-motor control.[5] These are extensive research areas by themselves; palpation intimately combines them. Successful investigations will require small pieces of the process to be analyzed separately. Then, a more complete picture can be built up from these individual studies.

Another important aspect of a remote palpation instrument is the mechanical design of the surgical instrument itself. It physically links the tactile sensor to the tactile display and allows the surgeon to scan the sensor over internal tissues. This couples the surgeon's finger motions (kinesthetic feedback) with small-scale shape information (cutaneous or tactile feedback). Ideally, the instrument should permit the surgeon to use hand and finger motions typically used in traditional palpation.[5] This would provide a delicate interface so the surgeon feels as if he or she is palpating the internal tissues directly. Developing a surgical instrument with a level of movement similar to a hand and fingers is a difficult task; it has remained challenging since, at best, only a few degrees of freedom are possible. The design challenge is further complicated by the mechanical constraints imposed by the minimally invasive surgical technique since the instrument must fit through a 5–15 mm diameter incision entry port.[5]

Design Specifications for Remote Palpation Instruments

Due to these challenges, design of a remote palpation instrument is complicated, and a variety of parameters must be taken into consideration. It is necessary to first determine the important aspects of palpation motions, and incorporate these into the design. Often, this will depend on the specific surgical application. From some researchers' observations, there are three main considerations:[5]

1. It must be determined if the tissue needs to be supported by the instrument. This is important for the overall structure because a grasper needs to be incorporated if the instrument must support the tissue. If there is sufficient support behind the tissue (either naturally or through the use of additional

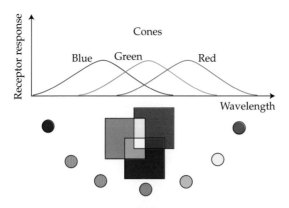

FIGURE 1.4 Colors created by the combination of cones responsible for three main wavelengths (colors).

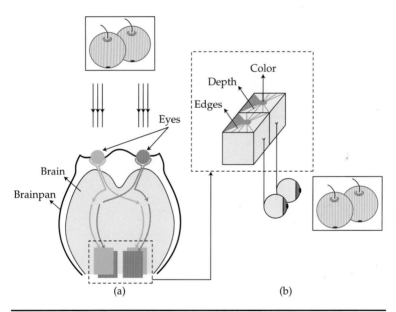

FIGURE 1.6 (a) Images seen by both eyes are interpreted in the left and right cortex. (b) Features such as color, depth, and edges are extracted by a different group of cells.

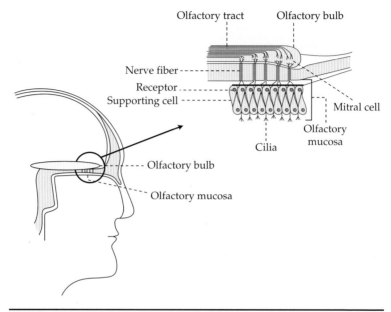

Figure 1.13 The graphical anatomy of the olfactory system.

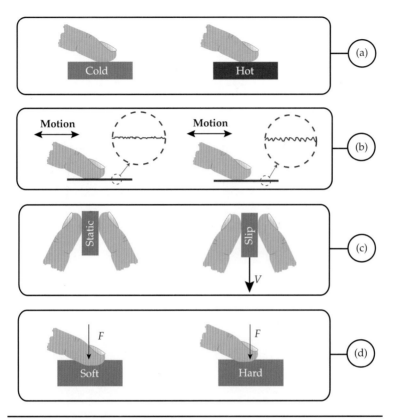

FIGURE 2.1 Some common features that can only be detected by our sense of touch: (*a*) thermal sense, (*b*) roughness of surface texture, (*c*) sense of slip, and (*d*) pressure and compliance detection.

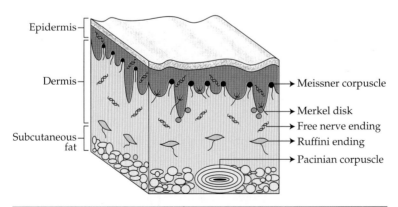

FIGURE 2.2 Layers of skin and location of its mechanoreceptors.

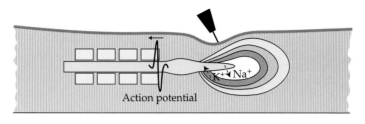

Figure 2.11 Generation of action potential and its propagation down the axon.

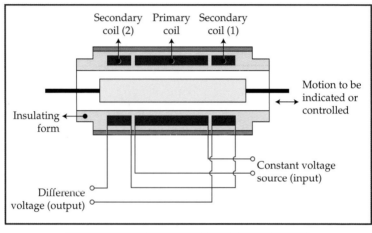

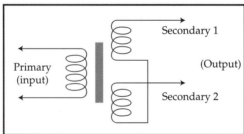

Figure 4.10 Typical structure of an LVDT.

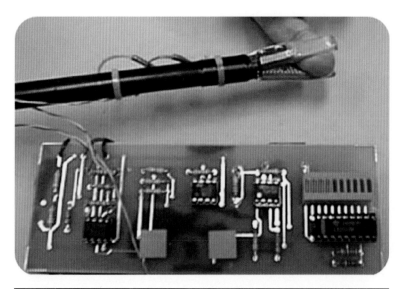

FIGURE 5.7 An endoscopic grasper and the associated electronic feedback system. The designed tactile sensor is mounted on the grasper. (Dargahi and Najarian, "An Endoscopic Force-Position Sensor Grasper with Minimum Sensors," *Canadian Journal of Electrical and Computer Engineering*, © 2003 IEEE.)

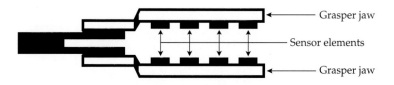

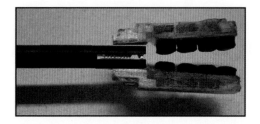

FIGURE 6.13 A commercial endoscopic grasper with four tactile sensors integrated with each jaw of the grasper. (Dargahi and Najarian, "Graphical Display of Tactile Sensing Data with Application in Minimally Invasive Surgery," *Canadian Journal of Electrical and Computer Engineering*, © 2007 IEEE.)

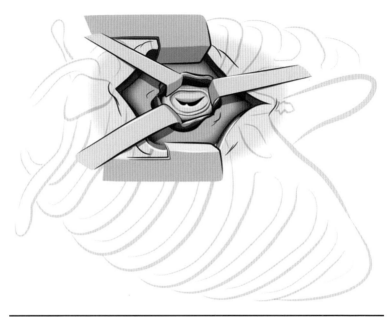

FIGURE 7.1 Schematic of an incision in open surgery. Metal fixtures are used to keep the incision open during the operation. (Photo courtesy of Intuitive Surgical, Inc., 2008.)

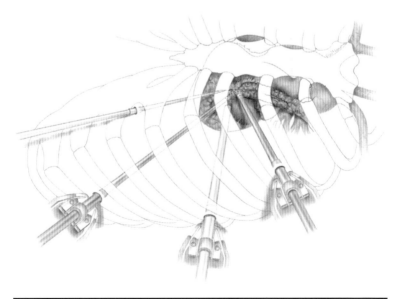

FIGURE 7.2 Surgical instruments are inserted into the patient's body through small incisions in MIS. (Photo courtesy of Intuitive Surgical, Inc., 2008.)

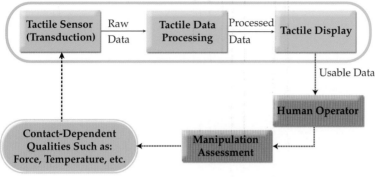

FIGURE 7.4 Working flowchart of a typical tactile sensing system for use in MIS. (Reprinted from *Mechatronics*, vol. 13, Eltaib and Hewit, "Tactile Sensing Technology for Minimal Access Surgery–A Review," pp. 1163–1177, Copyright © 2003, with permission from Elsevier.)

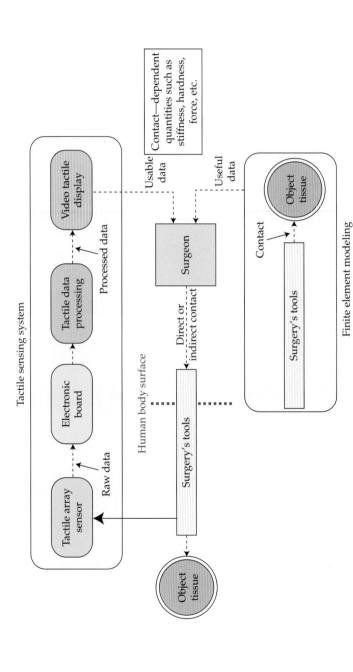

Figure 7.5 The components of a typical tactile sensing system and its role in surgery.

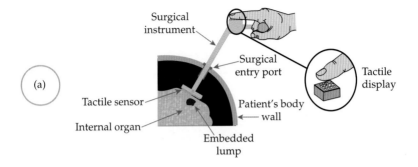

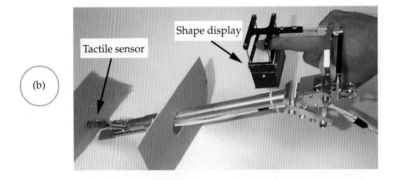

FIGURE 7.8 (a) Components of a surgical instrument for remote palpation and (b) a prototype of this instrument. (From Peine, used by permission, 1999.)

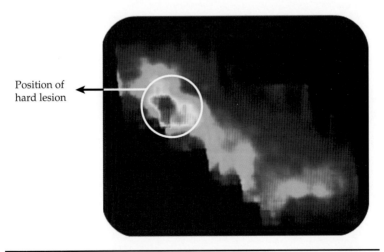

FIGURE 8.3 Sample tactile map. This map was obtained in short vertical swipes using the tactile imager shown in Fig. 8.2b. Note the hard lesion in red and yellow in the top left of the image. (From Galea, used by permission, 2004.[8])

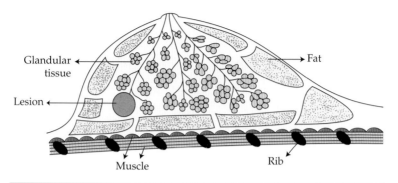

Figure 8.4 General anatomy of the human female breast, after puberty and before menopause.

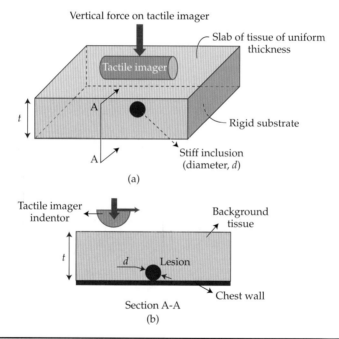

Figure 8.5 (a) The simplified model for study of parameter estimation from tactile information. (b) The centerline of the problem of interest. The arrows at the indentor indicate the direction of force (vertical arrow) and motion (horizontal arrow). (From Galea, used by permission, 2004.[8])

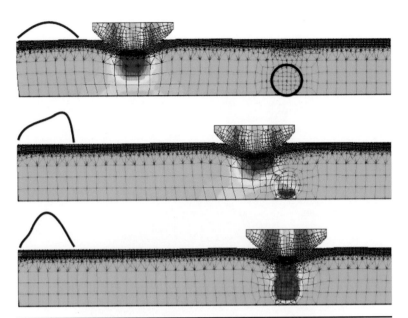

FIGURE 8.8 The finite element results of model in three conditions: indentor is far from the lesion, indentor is near the lesion, and indentor is placed just above the lesion. (From Galea, used by permission, 2004.[8])

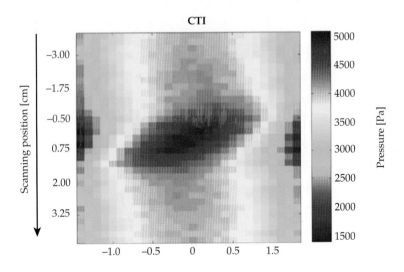

FIGURE 8.9 Composite tactile image based on data of finite element model. Each row approximates the pressure profile along the centerline of the indentor from a plane strain model. This CTI is a collection of the 33 pressure frames obtained every 2.5 mm over an 80 mm range with the lesion in the center. This CTI data is for a model with parameters of t = 25 mm, and d/t = 0.6, E_L/E_T = 50. (From Galea, used by permission, 2004.[8])

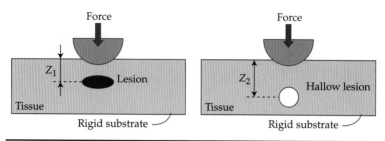

Figure 8.14 Two models with different lesions.

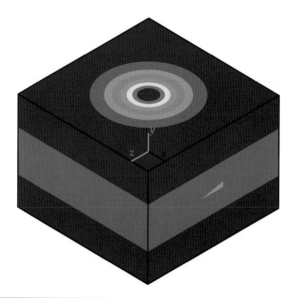

Figure 9.2 Tactile image for the code (sphere: l, 5 mm; t, 8 cm; d, 2 cm; h, 4 cm; Er, 30). ("Detection of Tumors Using Computational Tactile Sensing Approach," Hosseini et al.[6] Copyright © 2006 by John Wiley & Sons, Inc. Reproduced with permission of John Wiley & Sons, Inc.)

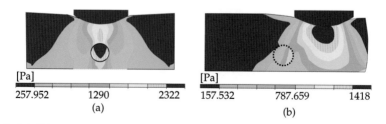

Figure 9.7 (a) Stress distribution for the code "Circle: l, 4 mm; t, 3 cm; d, 1 cm; h, 2 cm; Er, 30; x, +0 cm." (b) Stress distribution for the code "Circle: l, 4 mm; t, 3 cm; d, 1 cm; h, 2 cm; Er, 30; x, +2 cm."

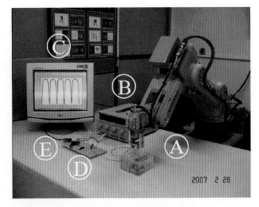

Figure 9.12 Experimental setup: (A) MOTOMAN robot, (B) Power supply, (C) Monitor or tactile display, (D) Electronic board, (E) Serial port, (F) Robot grippers, (G) Tactile probe, and (H) Phantom of tissue including tumor.

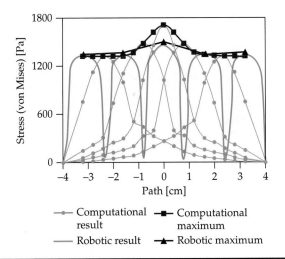

Figure 9.14 Operational comparison between computational and robotic results.

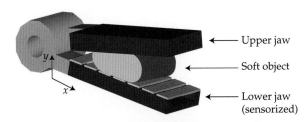

Figure 9.16 A view of the grasper with one active jaw equipped with an array of the seven sensing elements. (Sokhanvar, Ramezanifard, Dargahi, Packirisamy, "Graphical Rendering of Localized Lumps for MIS Applications," *Journal of Medical Devices*.[22] Copyright © 2007, Reproduced with permission of ASME International.)

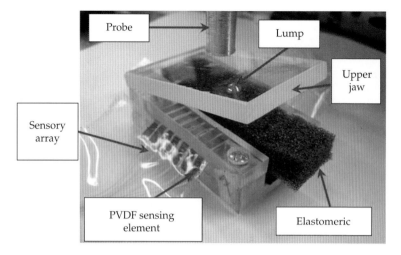

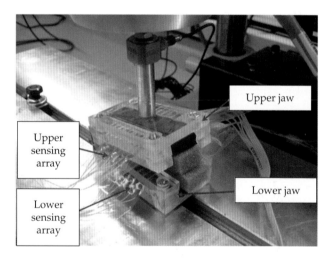

Figure 9.22 Photographs of the sensors under the test. (*a*) The sensor with one active jaw used for construction of one-dimension graphical images. (*b*) The sensor with two active jaws used for two-dimension graphical rendering of detected lumps. (Sokhanvar, Ramezanifard, Dargahi, Packirisamy, "Graphical Rendering of Localized Lumps for MIS Applications," *Journal of Medical Devices*.[22] Copyright © 2007, Reproduced with permission of ASME International.)

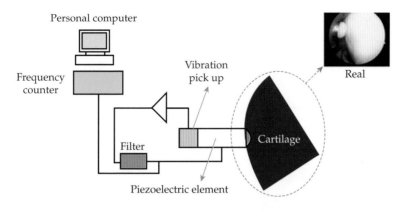

FIGURE 10.1 Tactile sensor structure. The change in resonance frequency when the sensor tip is attached to the tissue is a measure of the acoustic impedance of the tissue. (Reprinted from *Medical Engineering & Physics*, vol. 24, Uchio and et al. "Arthroscopic Assessment of Human Cartilage Stiffness of the Femoral Condyles and the Patella with a New Tactile Sensor," pp. 431–435, Copyright © 2002, with permission from Elsevier.)

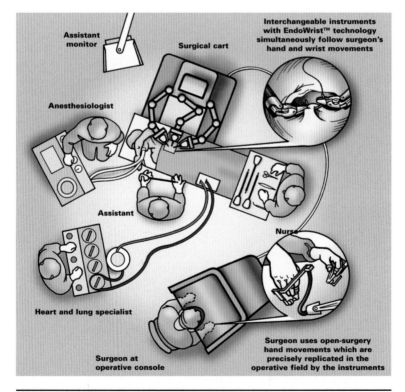

FIGURE 11.6 Schematic of performing suturing task by master-slave robot. (Photo courtesy of Intuitive Surgical, Inc., 2008.)

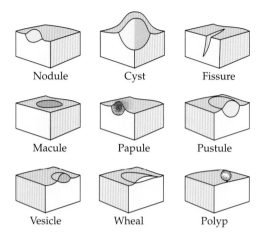

FIGURE 11.8 Some different types of skin lesions.

FIGURE 11.9 Appearance of anetoderma in (a) normal conditions and (b) when being touched. (Reprinted from White and Cox, *Diseases of the Skin:* Chapter 22, "Structural Disorders of the Skin and Disease of Subcutaneous Tissues; Structural Disorders of the Dermis, Collagen, and Elastic Tissue," Copyright © 2006, with permission from Elsevier.)

tools), direct probing on the surface is possible and a grasper is not needed.

2. It is important to control the contact between the tactile sensor and the tissue in a consistent way. This provides a regular input stimulus, or baseline, to the tactile sensory system. The tactile sensations measured by the sensor will be much easier to interpret due to the consistency of the baseline sensations, and this greatly simplifies the necessary signal processing. If the interaction is not well controlled, and the force and orientation of the finger varies significantly during the palpation motion, the baseline sensation will change.

3. Ergonomics and human factor issues are critical for extended use of the instrument. Typically, surgeons stand in one position and do not shift their weight for long periods of time during a MIS operation. They concentrate on a video monitor located slightly above eye level on the other side of the patient. Surgeons often feel fatigued after performing a minimally invasive procedure. This can result in an adverse effect such as applying a sudden excessive force to a scissors grip handle or suddenly moving the instrument in an undesired direction, causing damage to nearby tissues.

Analysis of Contact Force Between an Endoscopic Grasper Used in MIS and the Biological Tissues

Determining the contact pressure and force between the grasper and biological organ is very important; for this reason, we analyzed contact forces and stresses in our design of tactile sensors and surgical grasper instruments.

Endoscopic tools normally have a tooth-like grasper to easily grasp biological soft tissues. These surgical devices are of great use during manipulation tasks such as grasping internal organs, gentle load transfer during lifting, suturing and removing tissues. Some typical graspers are shown in Fig. 7.9.

To analyze the contact force between the grasper and the tissue, let us first assume the biological tissue is an ideal viscoelastic material presented by the Kelvin Model. This model includes a spring of modulus k_1 in series with a dashpot of viscosity η, which are parallel to another spring with modulus k_2. A schematic of this model is shown in Fig. 7.10.

For this material, the creep response to a step change in stress σ_0 is expressed as Eq. (7.1):

$$\varepsilon(t) = 0.5 J_1(t)\sigma_0 = \left\{ \frac{1}{k_1} + \frac{1}{k_2}\left[1 - \exp\left(\frac{t}{T_1}\right)\right] \right\}\sigma_0 \qquad (7.1)$$

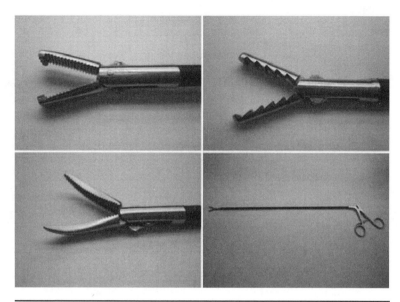

FIGURE 7.9 Typical endoscopic tools and graspers used in MIS. (From Bonakdar 2008, courtesy of Science Publications.)

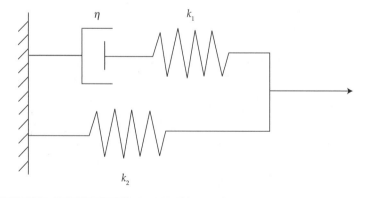

FIGURE 7.10 Kelvin model for viscoelasticity.

where $\varepsilon(t)$ is time dependent strain, $J_1(t)$ is the creep compliance and $T_1 = \eta/k_2$. The response to a step change of strain ε_0 can be written as Eq. (7.2):

$$s(t) = 2G(t)\varepsilon_0 = \frac{k_1}{k_1 + k_2}\left[k_2 - k_1 \exp\left(\frac{t}{T_1}\right)\right]\varepsilon_0 \qquad (7.2)$$

where $\sigma(t)$ is time dependent stress, $G(t)$ is the shear relaxation function or time dependent shear modulus and $T_2 = \eta/(k_1 + k_2)$.

Now, let us assume that a rigid and frictionless wedge as the grasper is indenting normally into the viscoelastic solid, as shown in Fig. 7.11. The contact pressure between the wedge and an elastic solid is expressed as Eq. (7.3):[10]

$$p(x) = \frac{E \times \cot\alpha}{\pi(1-v^2)} \cosh^{-1}\left(\frac{a}{x}\right) \qquad (7.3)$$

where E, v, and a are Young's modulus, Poisson's ratio, and a criterion for contact area respectively, as shown in Fig. 7.11.

Using Eq. (7.3), the changes of the contact pressure against x can be plotted. The plot is shown in Fig. 7.12. For an elastic object, the contact pressure gradually reduces at the tip of the wedge.

Since the biological tissues are considered to be ideal viscoelastic in our assumption for analysis, we should also apply viscoelastic theory to solve for the forces and stresses. We employ the linear theory of viscoelasticity, and we, additionally, consider the solid to be isotropic and homogeneous.

From these assumptions, contact pressure and area vary with time and, hence, the Poisson ratio is considered time independent and $E(t)$ is replaced by $2G(t)(1 + v)$. Consequently, Eq. (7.3) would become:

$$p(x,t) = \frac{2\cot\alpha}{\pi(1-v)} \int_0^t G(t-t') \frac{\partial \cosh^{-1}(a(t')/x)}{\partial t'} dt' \qquad (7.4)$$

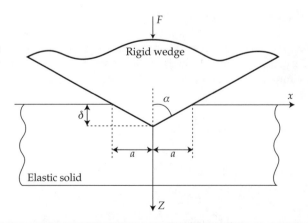

FIGURE 7.11 Rigid wedge in contact with a solid. (From Bonakdar 2007, courtesy of Science Publications.)

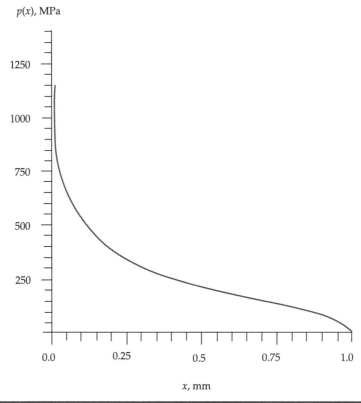

Figure 7.12 Variation of contact pressure along transverse direction (for $G = 235$ MPa, $a = 1$ mm, $\alpha = 60°$, $v = 0.5$). (From Bonakdar 2007, courtesy of Science Publications.)

The integral form of Eq. (7.4) can be interpreted as the linear superposition of small changes in $p(x, t)$ caused by infinitesimal step changes in contact area. Therefore, for delayed elasticity model defined by Eq. (7.1) and Eq. (7.2), for a constant value of contact area with incompressibility condition ($v = 0.5$), Eq. (7.4) can be rewritten as Eq. (7.5):

$$p(x,t) = \frac{4G(t)\cot\alpha}{\pi} \cosh^{-1}\left(\frac{a}{x}\right) \qquad (7.5)$$

For the elastic contact, the relationship between contact area, a, and force, F, is given by,

$$F = \int_0^t p(x)dx = \frac{2aG \times \cot\alpha}{(1-v)}$$

where $p(x)$ is taken from Eq. (7.3) and F is force per unit of length. Then,

$$a = \frac{F(1-v)}{2G \times \cot\alpha} \tag{7.6}$$

When the Poisson ratio is assumed to be 0.5 for a viscoelastic material, Eq. (7.6) can be written as Eq. (7.7):

$$a(t) = \frac{1}{4\cot\alpha} \int J_1(t-t') \frac{\partial F(t')}{\partial t'} dt \tag{7.7}$$

Under a step load, $a(t)$ becomes:

$$a(t) = \frac{1}{4\cot\alpha} J_1(t) F \tag{7.8}$$

By using Eq. (7.5) and Eq. (7.8), we can express the contact pressure as Eq. (7.9):

$$p(x,t) = \left\{ \frac{2k_1}{\pi(k_1+k_2)} \left[k_2 + k_1 \exp\left(-\frac{t}{T_2}\right) \right] \cot\alpha \right\} \cosh^{-1}\left(\frac{a}{x}\right) \tag{7.9}$$

and subsequently, $a(t)$ as Eq. (7.10):

$$a(t) = \frac{F}{2\cot\alpha} \left\{ \frac{1}{k_1} + \frac{1}{k_2}\left[1 - \exp\left(-\frac{t}{T_1}\right)\right] \right\} \tag{7.10}$$

Figure 7.13 shows the changes in contact pressure against time and location. As shown in the graph, at the constant value of the strain, contact pressure decreases rapidly and reduces towards a constant value which is the steady state.

In a grasping contact, by considering Eq. (7.8), the contact force per unit length on the top of the viscoelastic material for constant indentation area can be written as Eq. (7.11):

$$F(t) = \frac{2Nak_1}{(k_1+k_2)} \left[k_2 + k_1 \exp\left(-\frac{t}{T_2}\right) \right] \cot\alpha \tag{7.11}$$

where N is the number of teeth of the grasper. Figure 7.14 shows a schematic of a wedge teeth grasper.

Considering Eq. (7.10) for a constant indenting load, the creep of the contact area would be as Eq. (7.12):

$$a(t) = \frac{F}{2N \times \cot\alpha} \left\{ \frac{1}{k_1} + \frac{1}{k_2}\left[1 - \exp\left(-\frac{t}{T_1}\right)\right] \right\} \tag{7.12}$$

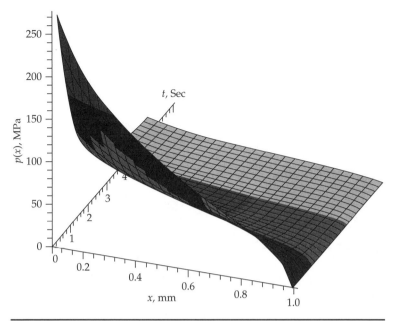

Figure 7.13 Variation of contact pressure in viscoelastic material against time and along transverse direction (for $k_1 = 235$ MPa, $k_2 = 26$ MPa, $\alpha = 60°$, $T_2 = 1$ Sec, $a = 1$ mm). (From Bonakdar 2007, courtesy of Science Publications.)

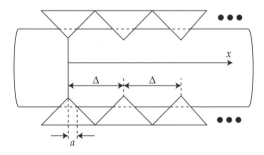

Figure 7.14 Schematic of grasper with wedge teeth. (From Bonakdar 2007, courtesy of Science Publications.)

This equation expresses the variation of contact area with force and time. As shown in Fig. 7.15, total force applied to the grasper decreases with time for a constant value of contact area.

From these characteristics, it should be noted that the contact area and, consequently, the contact depth will increase under a constant load. This is shown in the graph of Fig. 7.16. It is very important for a surgeon in MIS to realize this fact; otherwise, there is a possibility of causing damage to the organs.

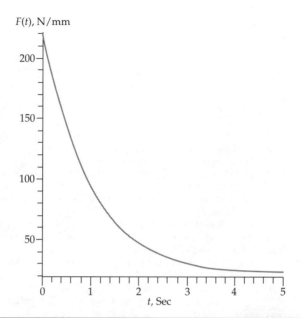

Figure 7.15 Decay of total force against time for the grasping contact with wedge teeth grasper (for k_1 = 235 MPa, k_2 = 26 MPa, α = 60°, T_2 = 1 Sec, a = 0.1 mm, N = 8). (From Bonakdar 2007, courtesy of Science Publications.)

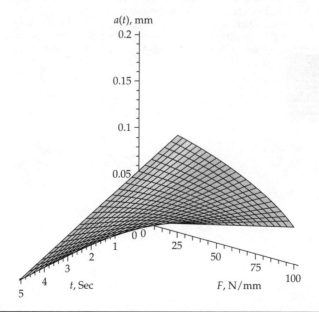

Figure 7.16 Increase of contact area with time in viscoelastic material against force (for T_1 = 10 sec, k_1 = 235 MPa, k_2 = 26 MPa, α = 60°, N = 8). (From Bonakdar 2007, courtesy of Science Publications.)

References

1. J. D. Wittchen, "A Study of Laparoscopic Instrument Mobility and Design," M.A.Sc. dissertation, University of Victoria, Canada, 1999. (Dissertations & Theses: A&I database, Publication No. AAT MQ40562.)
2. P. R. Nakka, "Design and Fabrication of a Micromachined Tactile Sensor for Endoscopic Graspers," M.A.Sc. dissertation, Concordia University, Canada, 2005. (Dissertations & Theses: A&I database, Publication No. AAT MR04423.)
3. A. Faraz, "Mechanisms and Robotic Extenders for Laparoscopic Surgery," Ph.D. dissertation, Simon Fraser University, Canada, 1998. (Dissertations & Theses: A&I database, Publication No. AAT NQ37699.)
4. H. Singh, "Design, Finite Element and Experimental Analysis of Piezoelectric Tactile Sensors for Endoscopic Graspers," M.A.Sc. dissertation, Concordia University, Canada, 2004. (Dissertations & Theses: A&I database, Publication No. AAT MQ94733.)
5. W. J. Peine, "Remote Palpation Instruments for Minimally Invasive Surgery," Ph.D. dissertation, Harvard University, United States, 1999. (Dissertations & Theses: A&I database, Publication No. AAT 9921526.)
6. H. Y. Yao, "Touch Magnifying Instrument Applied to Minimally Invasive Surgery," M.Eng. dissertation, McGill University, Canada, 2005. (Dissertations & Theses: A&I database, Publication No. AAT MR06595.)
7. P. L. Yen, "Palpation Sensitivity Analysis of Exploring Hard Objects Under Soft Tissue," *International Conference on Advanced Intelligent Mechatronics*, vol. 2, 2003, pp. 1102–1106.
8. R. D. Howe, W. J. Peine, D. A. Kantarinis, and J. S. Son, "Remote Palpation Technology," *IEEE Engineering in Medicine and Biology*, vol. 14, no. 3, 1995, pp. 318–323.
9. M. E. H. Eltaib, and J. R. Hewit, "Tactile Sensing Technology for Minimal Access Surgery–A Review," *Mechatronics*, vol. 13, 2003, pp. 1163–1177.
10. A. Bonakdar, J. Dargahi, and R. Bhat, "Investigations on the Grasping Contact Analysis of Biological Tissues with Applications in Minimally Invasive Surgery," *American Journal of Applied Sciences*, vol. 4, no. 12, 2007, pp. 1016–1023.

CHAPTER 8
Tactile Image Information

In this chapter, we will focus on tactile image information through palpation as another application of tactile sensing in medicine and surgery. As an example, we will discuss the idea by first explaining the importance of its application on the breast, which is one of the most common tissues used in palpating diagnosis; we will then develop the theoretical and engineering aspects of this method for applications in breast tissue.

8.1 Introduction to Palpation

In a typical physical examination, palpation is usually used. In this process, the healthcare practitioner uses the hands to feel an object in order to estimate its size, shape, firmness, or location.[1]

Palpation is an effective tool in a considerable number of medical procedures. Physicians usually use their sense of touch to examine patients. They conduct this procedure so that they can determine muscle tone, assess the size of the liver and spleen, and detect possible lumps in the breast. Palpation is also employed by surgeons. These healthcare professionals use this procedure to quickly determine the thickness of an artery wall, or even possibly to localize hidden tumors in various organs. Although palpation is used regularly, it has not been investigated thoroughly. It is believed that a detailed study of palpation can lead to two major benefits. The first benefit is that we can improve the current procedures and training; the second benefit is that we can develop more efficient devices for measuring, recording, and recreating tactile information during touch processes.[2]

From the mechanical and neuro-physiological points-of-view, palpation is an intricate process. We can employ contact mechanics analysis, nonlinear time-dependent material behavior, and large deformation solid mechanics in order to learn more about the mechanical stimulus to the fingers. The neuro-muscular control and human sensory process also contribute to the complexity of the analysis. Some research activities have been published on the correlation between mechanical stimulus to the finger, neural signals, and perception.[3] During palpation, the human finger is often in contact with surfaces of similar or greater compliance. Since both the finger and tissue deform, this changes the stimulus to the finger.

One of the most common palpation procedures is detecting and localizing a hard lump in soft tissue. This is the main objective of breast and prostate examinations, and an important part of several types of cancer surgery.

8.2 Taxonomy of Palpation

The mechanical design of surgical instruments is considered to be one of the most important aspects of remote palpation instrumentation.[4] Here, the tactile sensor is connected to a tactile display; in so doing, the surgeon can scan the sensor over internal tissues of the patient. This couples the surgeon's finger motions (kinesthetic feedback) with the small-scale shape information (cutaneous or tactile feedback). In an ideal design, the instrument would permit the surgeon to use hand and finger motions that are typically used in traditional palpation. As a result of this transparent interface, surgeons can feel as if they are directly palpating the internal tissues. As can be imagined, this task is very difficult to emulate.[4]

Palpation is dependent upon the excellent human ability to control the position and orientation of the hand and fingers.[4] During palpation of an organ, the surgeon's fingers scan consistently over the irregularly curved tissue. Here, the fingers are orientated normal to the surface. This keeps the finger pads in a uniform contact state with the tissue. The final result is that we have control of contact in both normal and tangential directions. Design and development of surgical instruments with this dexterity is a challenging issue.[4] In practice, it is only possible to incorporate a few degrees of freedom in the instrument. Additionally, the mechanical constraints imposed by MIS can further complicate the targeted design. In MIS, we use instruments that can be easily fitted through a 5–15 mm diameter entry port on the system.

Following the above-mentioned limitations, it is necessary to include the most important aspects of natural palpation motions into the design of the instrument. Different palpation motions are used for different tasks. Lateral motions, for instance, are used to sense texture, pressure is used to sense hardness, and unsupported holding is used for sensing weight.[4]

Researchers have developed a taxonomy that classifies the hand configurations used during the manipulation and grasping of various objects.[5] They divide the taxonomy based on two main types of grasps:

1. There are power grasps, where security and stability are emphasized.
2. There are precision grasps, where dexterity and sensitivity are important.

Following the above classification, each type of grasp is structured based on the complexity of the object's geometry, including whether it is flat, round, small, or long. Additionally, the requirements of the finger positions are also considered, such as prehensile using the thumb, contact with the palm, wrapping the fingers, the number of fingers, and so forth. Note that prehensile here means being able to grasp something or to take hold of things, especially by wrapping the fingers around them.

The finger motion has two components:[4]

1. It has normal forces/displacements into the tissue.
2. It has tangential motions caused by sliding or pushing along the surface of the tissue.

The type of tactile sensation associated with the task determines the amount of normal pressure required. Large forces are used to localize objects embedded deep in a soft tissue. To feel texture, however, only small forces are required. Following a contour of a bone, for example, requires scanning with the fingers over relatively large distances.[4] On the other hand, sensing tissue stiffness is primarily a local sensation that does not need any lateral motion. Therefore, it can be concluded that the extent of lateral motions depends on whether or not the tactile signals are spatially distributed. Surgeons select finger motions for a palpation procedure according to the information to be learned. Table 8.1 describes some of the most frequently used palpation motions, the corresponding parameter being sensed, and typical procedures that use these motions.

8.3 Palpation and Tactile Image

The principles of remote palpation and how surgeons can use them during MIS were described in the previous chapter. Palpation is used for the diagnosis of a wide range of tissues and organs in the body, such as the existence of a stone inside a gall bladder in open surgery. The organs shown in Fig. 8.1 are among those that palpation is used widely by surgeons in diagnosis through MIS.

Information for Mapping Tactile Imaging

For years, practitioners have used the human sense of touch as an established method for detecting pathologies of tumor tissue embedded under the surface of a soft organ. However, the human

Finger motion	Parameter being sensed	Example procedure
Steady applied pressure and no lateral motion	Time varying pressure or temperature	Locating arteries during dissection
Varying applied pressure and no lateral motion	Tissue stiffness	Detecting necrosis of the liver
Light applied pressure and fast lateral motion	Surface texture	Evaluating a rash on the skin
Heavy applied pressure and small lateral motion	Localization of hard occluded objects in soft tissue	Detecting lump during a clinical exam
Heavy applied pressure and large lateral motions	Contour following	Following a rib to determine location for an incision
Moderate applied pressure and differential motion between thumb and finger(s)	Function test	Sensing the thickness of an artery wall by rolling it between the fingers

Source: From Peine, used by permission, 1999.[4]

TABLE 8.1 Basic Palpation Motions

palpation as a method of pathology has been proven to be highly dependent on the skills of the individual practitioner.[6,7] In order to tackle this problem, scientists have focused on two issues; first, they develop a noninvasive and reliable method to measure the mechanical properties of soft tissue; secondly, they map embedded structures with different mechanical characteristics of original healthy tissue.

One of the most direct ways of implementing the clinical palpation is through tactile imaging.[8] Here, an array of passive pressure sensors is being used to map the surface pressures. The source of these surface pressures is from indenting the tactile imager into the surface of a soft material. The field of medical tactile imaging evolved for noninvasive recognition of tumor pathologies in biological soft tissues within the human body.[9,10] It is known that breast tissue is an easily accessible organ with pathologies that can be palpated or imaged using tactile imaging. Hence, the study of this organ has been one of the driving forces behind tactile imaging advancement.

Examples of tactile imagers[8] are shown in Fig. 8.2. Depending on the specific application for which the tactile imager was intended

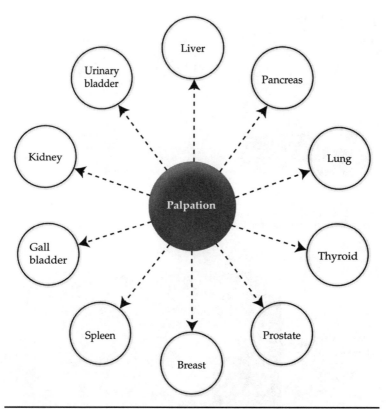

FIGURE 8.1 Tissues and organs typically diagnosed by palpation through MIS.

and designed, the exact size and shape of the tactile imager will vary. Despite this fact, the pressure arrays on tactile imagers used in studies are mounted on a cylindrical surface with a radius of curvature in only one direction. This is mainly due to manufacturing limitations.[8]

The purpose of the information gathered by tactile sensors in this system is to construct a two-dimensional tactile map, such as the one shown in Fig. 8.3. This provides a better understanding of the mechanical characteristics of the underlying tissues, such as its elastic properties.

Information about both the underlying material modulus and the geometry of the stiffness distribution can be obtained from the resulting tactile map. Tactile imaging can indicate the presence of stiff areas in soft tissues and present quantitative information about these areas of higher stiffness.[8,11]

In general, the tactile imaging method is more sensitive and accurate in detecting pathologies than clinical palpation. This is mainly because the tactile imaging method provides physicians with quantitative data about the scanned tissue or organ. Also, the resulting

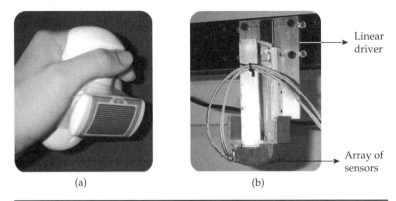

Figure 8.2 (a) Clinical tactile imager. (b) Laboratory tactile imager. (From Galea, used by permission, 2004.[8])

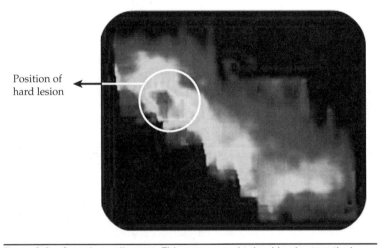

Figure 8.3 Sample tactile map. This map was obtained in short vertical swipes using the tactile imager shown in Fig. 8.2b. Note the hard lesion in red and yellow in the top left of the image. (From Galea, used by permission, 2004.[8]) (See also color insert.)

tactile map has shown to be reasonably repeatable. Therefore, it is capable of correcting possible problems reported for clinical palpation. Various medical fields have explored the application of tactile imaging. For instance, we can report liver and prostate tissues. In spite of these reports, this technique has been mostly applied to breast tissues.[8]

Based on the above discussion, the breast is an ideal candidate for full tactile imaging. Since it is an external organ, it has great social impact for patients. Additionally, the high prevalence of breast disease is good cause to investigate the role of tactile imaging in breast cancer screening and detection.[8] Before we proceed, let us first have a look at the anatomy of a healthy female breast.

Before menopause, the breast of a mature human female consists of glandular tissue (the mammary glands), adipose tissue (fat cells), and supporting tissue (stroma, mainly collagen and elastin fibers).[8] Its general anatomy is shown in Fig. 8.4.

The glandular tissue is located in a cone with its apex at the nipple. This tissue is also the site of milk production. The adipose tissue has layers under the skin and above the chest wall. In fact, this part is the greatest contributor to the general shape and size of the breast. The general appearance and motion of the breast is mainly related to the supporting tissue. This tissue is thin and light.

Menopause is the final stage of replacement of the glandular tissue by fat in a process called *involution*; during this process there is a decline or degeneration in the physiological function of the tissue. Involution of the breast usually begins at about thirty years of age. This process continues until menopause, during a period called *perimenopause*. It should be noted that there is a great variety in timing between individuals. Involution takes place as the glandular component atrophies and is replaced in discrete sections by adipose tissue. This will continue until the entire glandular component is replaced. Menopausal changes also happen to other components of the breast. As an example, there will be an increase in the volume of adipose tissue. At the same time, there will be a weakening process in the elastic and collagen fibers.

Depending on the relative age of the subject, the examination of the breast by palpation yields very different results.[8] A premenopausal breast has a high glandular tissue component under the superficial adipose tissue. This leads to an inhomogeneous texture. On the other hand, postmenopausal breasts are far more uniform. In these tissues, the adipose tissue is softer. Therefore, if one is searching for stiff pathologies, it is easier to palpate postmenopausal breasts.[8]

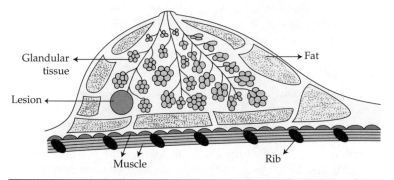

FIGURE 8.4 General anatomy of the human female breast, after puberty and before menopause. (See also color insert.)

Biological tissues are generally viscoelastic and strain-hardening. Following this, the tissues found in the human breast are no exception. However, some simplifications could be made to model and analyze the system numerically. With the breast supported against gravity and examined in compression, the glandular and adipose tissues almost entirely determine the local mechanical properties. Here, we observe linear elastic behavior for adipose tissue through strains of 15%. A Young's modulus of 15 ± 4 kPa is normally obtained in this regime.[8] Unlike adipose tissue, glandular tissue exhibits linear elastic behavior through strains of 6% at a Young's modulus of 45 ± 6 kPa.[8,12] Pathological tissues are stiffer. This means that they have a higher Young's modulus. The final result is that under palpation, we can feel the stiffer parts distinctively.[13,14] Also, if we are using tactile imaging, they can be viewed as a stiffer geometry of material.

It is estimated that in America, the probability that a woman could develop breast cancer in her lifetime is one in eight. In other words, breast cancer is responsible for one-third of all new cancer cases. This makes it second only to skin cancer as being the most common cancer in women. Statistical results have shown more than 178,000 new cases of breast cancer diagnosed among American women in 1998. More than 43,000 women die from this disease each year.[15] In the year of 2000, there were 182,800 new cases of breast cancer.[16] Hence, it is not surprising that breast cancer is the leading cause of cancer death of women next to lung cancer.[8] The statistics clearly show the importance of attaining a reliable and accurate method of detecting pathological tumor tissues embedded under the breast tissue.

Imaging Procedures for Breast Cancer

The most common and basic methods used for breast cancer screening have long been self breast examination, clinical breast examination, and mammography.[17] So far, these methods have been very useful in screening and detecting tumors. However, they have their own limitations. With regard to breast cancer, they fail to provide definitive and accurate diagnosis. Here, we provide a brief overview of each of these methods with their limitations. Then, we will introduce the advantages of the tactile imaging method.

Breast Self Exam

A *breast self examination* (BSE) is the visual inspection for any changes in the breast contour. This procedure is very useful because any pathology may affect the skin or ligaments of the breast. The final result is the distortion of the breast.[8] The other procedure is a manual palpation of the bulk of the breast to feel for the presence of any lump. Studies have reported that, on average, women only have a sensitivity score of 46%, and even with training about the proper technique and what to look for, this number would still only rise to

59%.[8] In addition, although the percentage of women who perform this test regularly is not known, it is postulated that this number is quite low.

Clinical Breast Exam

There is another type of breast examination which is similar to the self breast examination. It is called *clinical breast examination* or *CBE*, for short. A CBE, as with BSE, begins with the clinician observing the movement of the breasts with arm motion while the woman sits or stands. The woman then lays supine and the clinician manually palpates the breast. This is done by pressing gently with the pads of the fingers and continues until all the breast tissue is covered. Although most diagnoses from CBE are based on qualitative experience, the clinician tries to feel for nodules which exhibit the three D's. These are the nodules that are discrete, different, and dominant.[8]

Studies have revealed that even a trained clinician cannot feel a lump smaller than 3 mm in diameter and the diagnostic ability of the CBE is highly dependent on the clinician administering the test. Additionally, a proper CBE takes time and can make some patients and clinicians feel uncomfortable, so they are often not performed in full.

Mammography

The mammography technique is based on X-ray radiography. In this technique, only a very low dose of radiation is used. This happens because the breast consists solely of soft tissue. To obtain the maximum tissue spread and thinnest profile possible, the breast is compressed between two flat parallel plates. The path of the photons in the X-ray frequency range is as follows. First, they pass through the top plate, and then move through the breast tissue. The tissue is exposed on the photographic film in the bottom plate. On the X-ray image, a white region will be produced by regions of dense tissues.[8] The experience and expertise of the radiologist plays an important role in the interpretation of this mammography film. There are other disadvantages as well. For example, this procedure is very uncomfortable and is, sometimes, even painful. In certain cases, many lesions cannot be picked up by this technique. Additionally, benign densities often appear as tumor tissues on the film. Therefore, it is possible that this leads to false interpretation of the results. If this happens, the patient might even be asked to do more pathological tests.

Tactile Imaging and Breast Cancer Screening

Ideally, breast cancer screening involves information from both palpation and mammography.[8] However, to form a comprehensive source of information for cancer pathology, the information from various imaging modalities can rarely be integrated. This problem has to do with the manner in which each modality is performed. In practice, clinical breast examinations are performed by a clinician with the woman in a supine position. In mammography, however, the woman is standing. Since these data are obtained in different planes, it is difficult to combine them. To solve this, we can use tactile imaging to bridge the gap between these two pieces of information. This is done by providing a quantitative image of palpation. This image should be obtained in the same plane as a mammogram.[8]

Estimating of Lesion Parameters

In this section, we will use the breast as the target tissue to explain the theoretical basis of how typically tactile information is estimated by tactile imaging. Figure 8.5a shows a simplified overview of the geometry of an actual condition in which a stiff lesion is embedded in the soft biological tissue of a breast. Our target is to estimate the modulus in every point of the breast. To do this, we need to simplify the problem. Therefore, our focus will be on the centerline of the model as seen in Fig. 8.5b. We will consider the 2D model of Fig. 8.5b as a plane strain model. This model represents a long cylindrical lesion embedded in the tissue and indented by an infinite cylindrical scanhead.[8] Healthy breast tissue is approximated as a slab of material having a finite thickness.[14] It is assumed to be fixed to a flat, incompressible chest wall as seen in Fig. 8.5b. The lesion, which is stiffer than the background tissue, is also attached to the chest wall at one point. Additionally, we assume that the materials are linear elastic, isotropic, and perfectly bound to each other. The modulus of the tissue (background) and lesion are E_T and E_L, respectively.

Other assumptions in the modeling and theoretical analyses are as follows:

1. For a few centimeters near a breast lesion, the thickness of the tissue within the considered geometry of the model is constant. Since most breast lesions are found in the upper outer quadrant of the breast, where the soft tissue is relatively thin and flat, this assumption is not unreasonable.[8]
2. The lesion is considered to be round, since the clinical results are found to be in good correlation with the results from models with round lesions.[11]
3. The lesion is assumed to be attached to the fixed lesion, or substrate, below it.[8]
4. The contact between the tactile imager and tissue is frictionless. Provided that the surface of the tissue is well lubricated, this assumption is applicable to real tactile imaging situations.[11]

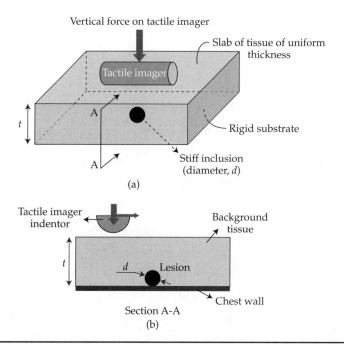

FIGURE 8.5 (a) The simplified model for study of parameter estimation from tactile information. (b) The centerline of the problem of interest. The arrows at the indentor indicate the direction of force (vertical arrow) and motion (horizontal arrow). (From Galea, used by permission, 2004.[8]) (See also color insert.)

5. According to what was described previously in the anatomy of breast and the tissues within, the breast tissue is considered to be an incompressible linear elastic material.[8,11,18] Again, this generally agrees with the actual behavior of biological tissues; they are comprised mainly of water and, so, can be considered incompressible. The elastic behavior and the magnitude of Young's elastic modulus for tissues comprising the breast have also been reported previously.

6. Any arbitrary element chosen within the geometry of the model undergoes plane strain. This assumption will be used to generate an inversion algorithm to estimate the parameters from tactile data.[8]

Analytical Solution

From these assumptions, the problem becomes the same as that presented in Fig. 8.6. Let us examine the analytical solution to the pressure profile when the indentor is directly over the lesion. Like any other mechanics problems, the solution must satisfy the equations of geometry, equilibrium, and the constitutive relations:[8]

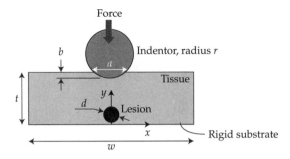

FIGURE 8.6 Developed model for palpation of breast tissue including lesion. (From Galea, used by permission, 2004.[8])

Geometry: The strain tensor ε is defined by its relationship with the displacement u of the body:

$$\varepsilon = \frac{1}{2}(\nabla u + \nabla u^T) \tag{8.1}$$

Equilibrium: Force equilibrium in continuum mechanics can be expressed by the following equation:

$$\nabla \cdot \sigma + f = 0 \tag{8.2}$$

where σ is the stress tensor and f is the local force tensor.

Establishing momentum equilibrium requires a symmetric stress tensor (i.e., the stress on the x-plane in the y-direction is equal to the stress on the y-plane in the x-direction, etc.):

$$\sigma = \sigma^T$$

Constitutive Relations: Continuous media can be characterized by the two-dimensional relation as Eq. (8.3):

$$\sigma_{ij} = C_{ijkl}\varepsilon_k \tag{8.3}$$

Here, σ and ε are 2-tensors, due to the symmetry of σ, as stated in the moment equilibrium equation.

The solution to Eq. (8.3) is subjected to the boundary conditions on all surfaces. They are: zero displacement on the bottom edge, a fixed displacement in the shape of the scanhead in the region under the tactile imager, zero stress on the sides, the areas of the top surface not in contact with the scanhead, and zero shear stress under the scanhead. The last boundary condition is based on the assumption of a frictionless contact. These can be mathematically expressed as:[8]

$u_1(x, y = 0) = u_2(x, y = 0) = 0$, for the bottom surface attached to a rigid substrate;

$u_2 (-a/2 \leq x \leq a/2, y = t) = -(r^2 - x^2)^{1/2} + r - b + t$, for displacement under indentor;

$\sigma_{11} (x = \pm w/2)$, for stress-free left and right edges;

$\sigma_{12} (x = \pm w/2)$, for shear stress-free left and right edges;

$\sigma_{22} (|x| > a/2, y = t) = 0$, for stress-free surface away from indentor;

$\sigma_{12} (y = t) = 0$, for no surface shear stress condition.

We can break the problem into small components, and use Finite Element Methods (FEM) to solve it. The FEM model is shown in Fig. 8.7. A finer mesh has been used in certain areas. This approach has a number of advantages. One benefit is that it allows for a more detailed representation of the large spatial variations in the stress field.[8]

The model is constructed with a plane strain assumption, with Poisson's ratio for each of the materials ranging from 0.45 to 0.49. This is because the glandular and fatty tissues in the breast are mostly water, an incompressible material. Simulation of the tactile imaging occurs by indenting the scanhead into the tissue and to the left of the inclusion. The vertical force is kept at a set value. By holding the force at this level, the scanhead is then moved laterally.[19]

Tactile Information from Finite Element Models

Figure 8.8 shows a sampling of the pressure frames obtained from the Finite Element Analysis. By pressing the indentor into the tissue far from the lesion, we can scan the surface. The details of the model are: tissue thickness $t = 20$ mm, lesion diameter $d = 10$ mm, background modulus $E_T = 15$ kPa, lesion modulus $E_L = 150$ kPa, and applied force = 8.0 N/mm. At left, we can see the pressure profile at the interface of the indentor and the tissue for each frame. Pressure frames were obtained every 2.5 mm for 40 mm to either side of the lesion.[8]

In order to use all of the tactile information of each frame, the frames were reassembled into a composite tactile image (CTI)[8] as

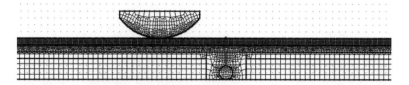

FIGURE 8.7 A typical finite element model used in this study. This model has 30 mm tissue thickness with an embedded lesion of 10 mm diameter. The tactile scanhead is modeled as a partial cylinder with infinite modulus. (From Galea, used by permission, 2004.[8])

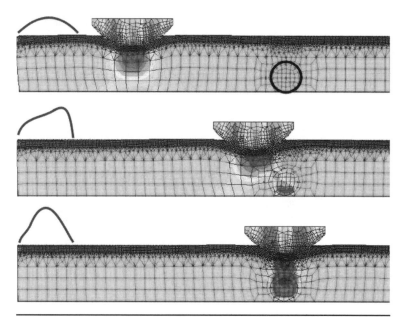

FIGURE 8.8 The finite element results of model in three conditions: indentor is far from the lesion, indentor is near the lesion, and indentor is placed just above the lesion. (From Galea, used by permission, 2004.[8]) (See also color insert.)

shown in Fig. 8.9. The pressure profile peak will shift to the right as the indentor approaches from the left of the lesion.

Inversion Algorithm

CTI data can be entered into a pressure vector P. Also, it is possible to construct a column vector G of the parameters of interest. Hence, the target will be finding the transformation matrix A that minimizes the error ε in Eq. (8.4):[8]

$$G = AP + \varepsilon \qquad (8.4)$$

for

$$G = \begin{bmatrix} E_T & E_L & t & d \end{bmatrix}^T \quad \text{and} \quad P = [CTI]_v$$

where the notation $[CTI]_v$ implies appending the rows of the CTI together into a row vector.[8]

The aforementioned system is a linear one. Here, the transformation matrix A is responsible for a linear transformation between the pressure data P and the parameters in G.

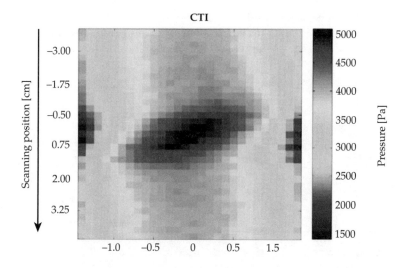

FIGURE 8.9 Composite tactile image based on data of finite element model. Each row approximates the pressure profile along the centerline of the indentor from a plane strain model. This CTI is a collection of the 33 pressure frames obtained every 2.5 mm over an 80 mm range with the lesion in the center. This CTI data is for a model with parameters of $t = 25$ mm, and $d/t = 0.6$, $E_L/E_T = 50$. (From Galea, used by permission, 2004.[8]) (See also color insert.)

If we move far away from the lesion, the parameters that will affect the surface pressure are the background modulus E_T and thickness t. It is assumed that the total scanhead force F can be described by $F = PA_s$. Here, A_s is the area of the entire scanhead and P is the representative pressure recorded. As shown in Fig. 8.10, we can approximate the tissue directly under the scanhead as a linear spring. Therefore, F can be expressed by Eq. (8.5):[8]

$$F = \frac{E_T A_s}{t} \Delta t \qquad (8.5)$$

where Δt is the indentation distance into the tissue. Then:

$$P = \frac{E_T}{t} \Delta t$$

We see that the pressure information P is related directly to the background modulus E_T. Also, it is observed that P is inversely related to the tissue thickness t. Therefore, we can conclude that the indentation distance depends on the force applied.[8] By plotting the maximum background pressure versus the linear parameters E_T and $1/t$, it is evident that there is a linear relationship between these parameters and the peak pressure obtained in the absence of a lesion. The plots are shown in Fig. 8.11.

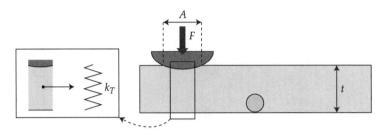

Figure 8.10 The mechanical behavior of the tissue under the sensor can be approximated as a spring. (From Galea, used by permission, 2004.[8])

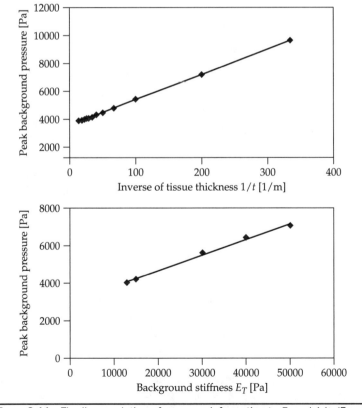

Figure 8.11 The linear relation of pressure information to E_T and $1/t$. (From Galea, used by permission, 2004.[8])

The relations become a bit more complicated when the indentor goes over the lesion. By considering the lesion, and the background tissue above it, to be represented by a linear spring each, then the tissue through the lesion can be modeled as two springs in series.[8] The spring constant for the spring through the lesion is E_L/d. The same constant through the tissue is $E_T/(t-d)$, as seen in Fig. 8.12. The total spring constant k seen by the tactile imager directly over the lesion can be expressed by Eq. (8.6):

$$k = \left(\frac{t-d}{E_T} + \frac{d}{E_L}\right)^{-1} = \frac{E_L E_T}{E_L(t-d) + E_T d} \quad (8.6)$$

where E_T and E_L are background modulus and lesion modulus, respectively.

Many cases of breast pathology are similar to the models in which a lesion is resting on a rigid substrate. A model which did not have a lesion attached to the substrate is also useful. They can represent breast pathology in women with large breasts.[8] This model is illustrated in Fig. 8.13.

In such cases, a fifth parameter, z, the depth of the lesion[8,11] will be taken into account for the analysis, which must also be estimated from the tactile information. Similar to the previous case, we can study the estimation method on results obtained from finite element models constructed with the same assumptions.

By using tactile imaging techniques, features that are critical for physicians in their diagnosis of cancerous tumors can be estimated. These features include:[11]

- Hardness (elasticity modulus) of lesion
- Size of lesion
- Shape of lesion, as in Fig. 8.14
- Depth of lesion, as in Fig. 8.14

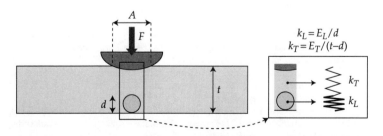

Figure 8.12 Representation of the tissue and the lesion as two springs in series. (From Galea, used by permission, 2004.[8])

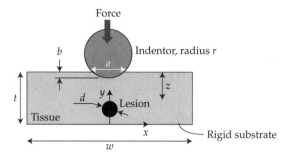

FIGURE 8.13 Model for parameter estimation from tactile information. The lesion is allowed to float in the tissue, and is no longer attached to the rigid substrate. The other parameters and assumptions remain the same as in previous model. (From Galea, used by permission, 2004.[8])

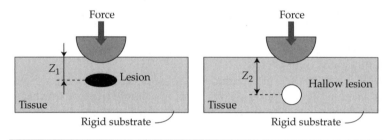

FIGURE 8.14 Two models with different lesions. (See also color insert.)

The tactile imaging method has proved to be of great use as a diagnostic tool that supplies physicians with valuable and accurate quantitative data; so far, no other method has, hitherto, been capable of providing this type of valuable data.

References

1. http://en.wikipedia.org/wiki/palpation
2. R. D. Howe, W. J. Peine, D. A. Kantarinis, and J. S. Son, "Remote Palpation Technology," *IEEE Engineering in Medicine and Biology*, vol. 14, no. 3, 1995, pp. 318–323.
3. W. J. Peine and R. D. Howe, "Do Humans Sense Finger Deformation or Distributed Pressure to Detect Lumps in Soft Tissue," *Proceedings of The ASME Dynamic Systems and Control Division*, vol. 64, 1998, pp. 273–278.
4. W. J. Peine, "Remote Palpation Instruments for Minimally Invasive Surgery," Ph.D. dissertation, Harvard University, United States, 1999. (Dissertations & Theses: A&I database, publication No. AAT 9921526.)
5. M. R. Cutkosky and R. D. Howe, "Human Grasp Choice and Robotic Grasp Analysis," *Dextrous Robot Hands*, S. T. Venkataraman and T. Iberall (eds.), Springer-Verlag, New York, 1990.
6. M. F. Evans, "False-Positive Results in Breast Cancer Screening," *Canadian Family Physician*, vol. 45, 1999, pp. 73–74.

7. I. Jatoi, "Breast Cancer Screening," Chapman and Hall, New York, 1997.
8. Anna M. Galea, "Mapping Tactile Imaging Information: Parameter Estimation and Deformable Registration," Ph.D. dissertation, Harvard University, United States, 2004. (Dissertations & Theses: A&I database.)
9. E. H. Frei, B. D. Sollish, S. Yerushalmi, S. B. Lang, and M. Moshitzky, "Instrument for Viscoelastic Measurement," USA Patent #4, 144, 877, March 20, 1979.
10. A. I. West, D. Krag, J. B. Weinstein, and N. Dewagan, "Obtaining Images of Structures in Bodily Tissue," USA Patent # 6500119, December 31, 2002.
11. M. Hosseini, S. Najarian, S. Motaghinasab, and J. Dargahi, "Detection of Tumors Using Computational Tactile Sensing Approach," *International Journal of Medical Robotics and Computer Assisted Surgery*, vol. 2, no. 4, 2006, pp. 333–340.
12. T. A. Krouskop, et al., "Elastic Modulus of Breast and Prostate Tissues under Compression," *Ultrasonic Imaging*, vol. 20, 1998, pp. 260–274.
13. S. Najarian, J. Dargahi, and V. Mirjalili, "Detecting Embedded Objects Using Haptics with Applications in Artificial Palpation of Tumors," *Sensors & Materials*, vol. 18, no. 4, 2006, pp. 215–229.
14. S. M. Hosseini, S. Najarian, S. Motaghinasab, A. Tavakoli Golpaygani, and S. Torabi, "Prediction of Tumor Existence in the Virtual Soft Tissue by Using Tactile Tumor Detector," *American Journal of Applied Sciences*, vol. 5, no. 5, 2008, pp. 483–489.
15. S. H. Landis, T. Murray, S. Bolden, and P. A. Wingo, "Cancer Statistics, 1998." *CA—A Cancer Journal for Clinicians*, vol. 48, no. 1, (January/February 1998), pp. 6–29.
16. R. T. Greenlee, T. Murray, S. Bolden, and P. A. Wingo, "Cancer Statistics, 2000." *CA—A Cancer Journal for Clinicians*, vol. 50, no. 1, 2000, pp. 7–33.
17. B. Matic, "Sensor Technology for the Breast Examination Training Instrument," M.A.Sc. dissertation, West Virginia University, United States, 2001. (Dissertations & Theses: A&I database, publication no. AAT 1407677.)
18. A. Abouei Mehrizi, S. Najarian, M. Moini, and F. Tabatabai Ghomshe, "Tactile Distinction of an Artery and a Tumor in a Soft Tissue by Finite Element Method," *American Journal of Applied Sciences*, vol. 5, no. 2, 2008, pp. 83–88.
19. G. Weber, "Using Tactile Images to Differentiate Breast Tissue Types," Undergraduate thesis in engineering, Harvard University, U.S.A., 2000. (http://www.griffinweber.com/thesis/).

CHAPTER 9
Application and Recent Developments of Tactile Sensing in Tumor Detection

9.1 Introduction

Tactile sensing is a new approach to detect embedded objects in biological tissues. This approach, in comparison to the other methods, is noninvasive. Dargahi and Najarian[1,2] have carried out two comprehensive surveys on this field of research. Their first survey reviewed human tactile perception as a standard for this technology; then, they evaluated advances in tactile sensors and their impact on robotics applications. Hosseini et al.[3,4] proved the reliability and accuracy of the artificial tactile sensing approach for the detection of tumors in biological tissues, using the finite element method (FEM); Najarian et al.[5] proposed a new analytical method that can be used as a predictive tool for determining both the stiffness and certain geometrical details of embedded objects.

In this chapter, we explain recent developments in detecting embedded objects, such as a tumor, in a simulated biological tissue by simulating palpation. The use of a finite element method, an analytical method, and experimental models are suggested for this application.

9.2 Detection of Tumors Using a Computational Tactile Sensing Method

In many diagnostic tests, such as clinical breast examinations, doctors usually examine the patient's body with their finger tips and hand palm to obtain information on conditions inside the body, including

Chapter 9

the presence of a tumor and its precise characteristics. In practice, when a woman visits her physician, part of the physical examination often includes a clinician attempting to palpate the patient for any lumps or changes in the breast tissue that could indicate the presence of a tumor. This method, however, only gives the physician a vague sense of what is actually under the skin due to the lack of any precise and consistent measuring approach. Of course, if a lump is found through palpation, typically all that can be documented is its rough location on the breast and an approximate estimate of its size. In addition, if the same patient visits other clinicians, she will, undoubtedly, receive a different description of her status, again due to inconsistent methodology. In this part of the book, we develop the concept of a consistent approach which requires neither the dexterity of a reputable physician nor any penetration/invasive procedure as is encountered during an X-ray in mammography.

Three-dimensional analysis leads to a novel method of predicting the characteristics of a tumor; it can be directly incorporated with tactile sensing in artificial palpation, helping surgeons in non-invasive procedures. Accordingly, a new method for determining the existence of an embedded object in biological tissue is presented. In this method, the indications of an existing tumor that appear on the surface of the tissue are determined. The use of FEM is needed for providing properties such as the shape, size, depth, and location of a tumor. Figure 9.1 shows a transverse section of a tissue (cube) and tumor (sphere), labeled with input parameters. In the following paragraph, we describe this method.

The aim is to investigate the effects of an object embedded in a biological tissue associated with the application of mechanical loading on the tissue. The stress distribution solution is called *tactile image*. Here, a graph is extracted plotting the stress magnitude on a specific axis and is called the *stress graph*. Then, the changes of the tissue response to variations of the input parameters are investigated as *tactile maps*. A code with the general form $(s: l; t; d; h; Er)$ is defined, showing the value of each input parameter, and the code (sphere: l, 5 mm; t, 5 cm; d, 2 cm;

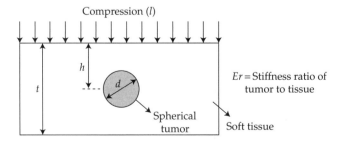

FIGURE 9.1 Transverse of a tissue (cube) and tumor (sphere), labeled with input parameters.

h, 2.5 cm; Er, 30) is dedicated as that of the base model. Any new model is constructed by changing just one input parameter of the base model. Therefore, it is possible to analyze the variation of tissue response vs. changes of any of the input parameters. The input parameters are:

1. Tissue loading (compression of the upper surface of the cube), l
2. Tissue thickness, t
3. Tumor diameter, d
4. Tumor depth, h
5. Stiffness ratio of tumor to tissue, Er
6. Tumor shape, s

From a predetermined range of input parameters, the following results can be elicited from tactile images and stress graphs:[6]

- Appearance of the effects of an embedded object on the surface in tactile images: The appearance of the symptoms of the tumor on the surface of the tissue is the most fundamental result that confirms the accuracy and reliability of the artificial tactile method. The tactile image for the code (sphere: l, 5 mm; t, 8 cm; d, 2 cm; h, 4 cm; Er, 30) is illustrated in Fig. 9.2 and

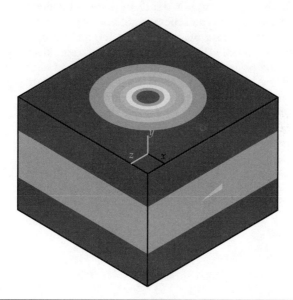

Figure 9.2 Tactile image for the code (sphere: l, 5 mm; t, 8 cm; d, 2 cm; h, 4 cm; Er, 30). ("Detection of Tumors Using Computational Tactile Sensing Approach," Hosseini et al.[6] Copyright © 2006 by John Wiley & Sons, Inc. Reproduced with permission of John Wiley & Sons, Inc.) (See also color insert.)

demonstrates the application of compression to tissue which contains an embedded object. This will cause a nonuniform stress distribution to be produced at the contact surface.

- Determining the tumor shape and size with respect to the stress distribution produced on the surface: In tactile images, stress contours produced on the surface of the tissue indicate the shape and size of the tumor; hence, if the stress contours on the surface expand circularly, then the tumor will be spherical. If the tumor is oval, the stress contours are enlarged elliptically. In addition to predicting the shape of the tumor, it is possible to estimate its size. The larger the contours, the bigger the tumor.

- Appearance of an overshoot in stress graphs: The stress graph is taken on a path, from left to right, defined by a straight line in the middle of the upper surface of the cube. The stress graphs that are taken along the defined path show an increase (overshoot) in the amount of stress that demonstrates the existence of a tumor. Figure 9.3 illustrates the stress graph of the base model. This overshoot of the graph not only confirms the tumor's existence, but also locates its exact position. This is because the position of the summit in each graph coincides with the center of the tumor in the tissue.

- Predicting the tumor depth by using changes occurring in stress graphs: The schematic shapes of all stress graphs are similar to those of the base model, including an overshoot in the middle (because of the tumor located in the middle of the tissue) and two decreasing parts in each side of the overshoot. However, variations of some input parameters can change this default shape of the graphs. In an inversion

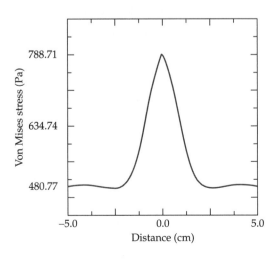

FIGURE 9.3
Stress graph for the base model (sphere: *l*, 2 cm; *h*, 2.5 cm; *Er*, 30).

analysis of the graphs, it is possible to determine the related parameters of the tumor. Tumor depth can be one of these parameters.

Figure 9.4 shows the stress graphs related to different values of parameter h in a determined range. It can be concluded that by increasing the tumor depth, the decreasing parts of the graphs become smaller and, eventually, disappear. Therefore, the presence or absence of the decreasing parts of the graphs indicates the low and high depth of the tumor, respectively.

A significant and discriminating aspect of this method is the three-dimensional analysis of the tissue and tumor.[1] This is true because performing three-dimensional modeling is the only way to observe and examine the contours produced on the surface of the tissue that, in turn, are used to determine the shape of the tumor. Moreover, it can be concluded that by referring to tactile maps the characteristics of the tumor can be anticipated.

Another important feature of this approach is the benefits of the results for surgeons in their surgical operations. The important immediate application is that, using artificial tactile sensing, the surgeon can easily obtain adequate information about the progress of the tumor using a noninvasive procedure in contrast to that which occurs during mammography, ultrasonography and MRI.[6] The other practical use of these results can be extended to the field of MIS, where the sense of touch is absolutely necessary for the surgeons to compensate for their inability to palpate the operation sites. Tactile images and/or maps add more information to the unfavorable visualizations provided for surgeons in MIS.

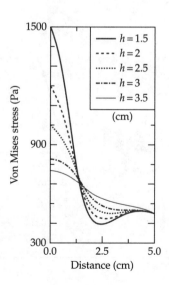

FIGURE 9.4 Disappearance of decreasing parts of the stress graphs by increasing the value of parameter h. ("Detection of Tumors Using Computational Tactile Sensing Approach," Hosseini et al.[6] Copyright © 2006 by John Wiley & Sons, Inc. Reproduced with permission of John Wiley & Sons, Inc.)

Figure 9.5 shows a simplified model of the problem with the related input parameters. Accordingly, a code with the general form of "s: l; t; d; h; Er; x" was defined, showing the value of each input parameter for every model. For example, the model with the code "Circle: 4 mm, 3 cm, 1 cm, 2 cm, 30, +1 cm" means that: the tumor has a circular shape; the compression of the tactile sensor in the upper surface of the tissue is 4 mm; the thickness of the tissue is 3 cm; the diameter of the circular tumor is 1 cm; the center of the tumor is 2 cm away from the surface of contact; the tumor is 30 times stiffer than the surrounding tissue; and the distance between the center of the sensor and the center of the tissue is 1 cm in the right side of the tumor.

Figure 9.6 shows a number of different stress graphs obtained by changing the sensor position (Input Parameter x). If the maximum points of these stress graphs are connected together, the peak in the

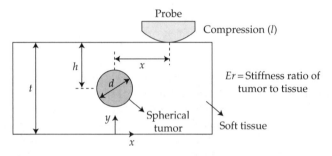

Figure 9.5 Simplified model of the problem labeled with input parameters.

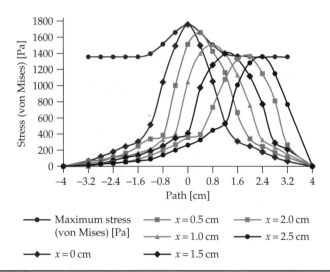

Figure 9.6 Variations of stress graphs.

resulted graph not only confirms the presence of a stiffer object inside the tissue, but also locates the exact position of the tumor.

Figure 9.7 shows the stress distribution for two different sensor positions. Since the problem is symmetrical, the function of the sensor is considered only on one side of the tumor. Figure 9.7 illustrates the stress distribution on the tissue surface. In the artificial tactile sensing approach, the basis of collecting information is the contact position (tissue surface); therefore, the precise investigation of this event is of crucial importance. To carefully analyze what is happening on the surface, a path is defined exactly on the upper surface of the rectangle and the stress taken on this path is depicted in a distinct graph called "Stress Graph." Figures 9.7a and 9.7b show stress distribution for the code "Circle: *l*, 4 mm; *t*, 3 cm; *d*, 1 cm; *h*, 2 cm; *Er*, 30; *x*, +0 cm," and "Circle: *l*, 4 mm; *t*, 3 cm; *d*, 1 cm; *h*, 2 cm; *Er*, 30; *x*, +2 cm," respectively.

9.3 Application of Artificial Neural Networks for the Estimation of Tumor Characteristics in Biological Tissues

Computerized methods have recently shown great potential in assisting radiologists in visual diagnosis of lesions, by providing them with a second opinion about the degree of malignancy of a lesion.[7-11] Here, inverse problems, which are very popular in engineering and science, play a crucial role in image restoration,[12] physiological systems,[13] and the analysis of DNA segmentation.[14] These are inverse problems because the input (the cause) of a system is reconstructed from output measurements (the effects).[15]

Although it is predicted that the results of the forward approach can show the existence of tumors, they do not provide the physicians with detailed information on the tumor characteristics. Among these characteristics, the depth, size, and stiffness of the predicted tumor are the most important, since this related information helps physicians to choose an appropriate treatment approach. One technique for solving inverse problems, using an artificial neural network (ANN), is the use of an inverse method on forward results. Figure 9.8 shows

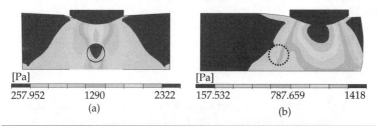

FIGURE 9.7 (a) Stress distribution for the code "Circle: *l*, 4 mm; *t*, 3 cm; *d*, 1 cm; *h*, 2 cm; *Er*, 30; *x*, +0 cm." (b) Stress distribution for the code "Circle: *l*, 4 mm; *t*, 3 cm; *d*, 1 cm; *h*, 2 cm; *Er*, 30; *x*, +2 cm." (See also color insert.)

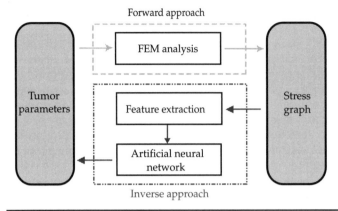

FIGURE 9.8 Flowchart of forward and inverse approaches. ("Application of Artificial Neural Networks for Estimation of Tumor Characteristics in Biological Tissues," Hosseini et al.[19] Copyright © 2007 by John Wiley & Sons, Inc. Reproduced with permission of John Wiley & Sons, Inc.)

a flowchart of this problem. ANNs are characterized in principle by a network topology, a connection pattern, neural activation properties, a training strategy, and the ability to process data. The rapid development of ANN technology in recent years has led to an entirely new approach on the solution of tactile data processing, usually applied to forward results of stress analysis.[16,17] In fact, neural networks are artificial intelligence systems mimicking the biological processes of a human brain by using nonlinear processing units to simulate the functions of biological neurons.[18]

Neural networks have been employed to determine the tumor characteristics, where stress graphs were the inputs of the neural network and the tumor characteristics were the desired outputs.

In this approach, to train the network effectively, a feature extraction step had to be undertaken. The position of this step is illustrated in Fig. 9.8 and involved examining the data collected through the forward results. In selecting these features, all of the obtained stress graphs were carefully examined. To determine which features were most important in characterizing these data, we focused on the stress graphs to extract some features that best corresponded to the desired tumor characteristics: diameter (d), depth (h), and tumor/tissue stiffness ratio (Er). Five features extracted from the stress graphs were as follows:[19]

1. The maximum stress point
2. The stress value corresponding to one-fifth of the distance between the peak and the beginning of the stress graph
3. The stress value corresponding to two-fifths of the distance between the peak and the beginning of the stress graph

4. The beginning stress
5. The transverse value (fatness or narrowness) of the stress graphs at 50% of the difference between the maximum stress point and the beginning stress of the stress graphs

Figure 9.9 shows these features in a typical stress graph. Using these five features, we have converted and reduced the total stress graphs into five most effective numbers, which are the inputs of the neural network. For example, Fig. 9.10 shows stress graphs for different values of tumor diameter (d) when other tumor characteristics are kept constant.

There is a nonlinear correlation between the tumor characteristics and their effects on the extracted features. In general, reliable estimation of tumor stiffness is obtained when the depth of the tumor is small, as shown in Fig. 9.11.

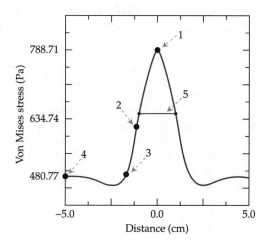

Figure 9.9
Extracted features in the stress graph. ("Application of Artificial Neural Networks for Estimation of Tumor Characteristics in Biological Tissues," Hosseini et al.[19] Copyright © 2007 by John Wiley & Sons, Inc. Reproduced with permission of John Wiley & Sons, Inc.)

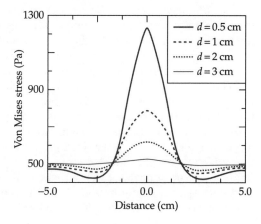

Figure 9.10
Stress graphs for different values of tumor diameter. ("Application of Artificial Neural Networks for Estimation of Tumor Characteristics in Biological Tissues," Hosseini et al.[19] Copyright © 2007 by John Wiley & Sons, Inc. Reproduced with permission of John Wiley & Sons, Inc.)

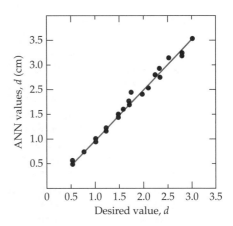

FIGURE 9.11
Comparison of tumor diameter estimation calculated by ANN and desired values: diagonal representation. ("Application of Artificial Neural Networks for Estimation of Tumor Characteristics in Biological Tissues," Hosseini et al.[19] Copyright © 2007 by John Wiley & Sons, Inc. Reproduced with permission of John Wiley & Sons, Inc.)

9.4 Prediction of Tumor Existence in the Virtual Soft Tissue by Using Tactile Tumor Detector

As previously described, finite element analysis provided properties such as the shape, depth, and location of the tumor, which are the most important parameters for physicians to distinguish the correct condition of the patients. Several different cases were created and solved by the ANSYS software, while tactile images and stress graphs were extracted.[6] These results clearly showed the existence of the tumor in the tissue. This part of the book presented the innovation of an artificial tactile sensing system called the *Tactile Tumor Detector*,[20] which mainly consisted of three components: the tactile probe, the tactile data processor, and the tactile display. Having performed a number of experiments, there was good correlation between the numerical and experimental results obtained. In addition to anticipating the presence of a tumor in the tissue and locating its exact site, the experimental results also helped the user to predict the depth of the tumor inside the tissue.

Tactile Tumor Detector Device Design: In order to validate the results obtained by the numerical analysis of the finite element method, a system named Tactile Tumor Detector was built, which works according to Fig. 7.4 in Chapter 7. This figure shows the main components and data circulation in this device.

The main part of the tactile probe, which can be easily handled, is a *force sensing resistor* (*FSR*) supplied by Interlink Electronics, Camarillo, CA. FSRs are polymer thick film (PTF) devices which display a decrease in resistance with an increase in the force applied to the active surface. Its force sensitivity is optimized for application in human touch control of electronic devices. This quality makes this kind of sensor appropriate for use in tactile sensory systems that deal with special configuration of forces. The sensor is placed on a dome-like probe to provide effective contacts.

A number of models were constructed to simulate a tumor inside the tissue by using a typical gel containing an alien object. Then, the model was scanned by the tactile probe on a straight line, passing over the tumor on the surface of the model. Figure 9.12 shows the tactile sensing system with one of the constructed models.

Figure 9.13 shows the result obtained by performing an experiment on one of the constructed models with a spherical simulated tumor. This simulated tumor is placed near the surface of contact.

The horizontal axis measures the straight line that the tactile probe has followed. Intentionally, it is labeled from −5 cm to 5 cm to be similar to the stress graph depicted in Fig. 9.3. The vertical axis is the processed outputs of the FSR, which is proportional to the force applied by the surface of the constructed models to the active area of the FSR.

In Fig. 9.14, the operation of the Tactile Tumor Detector is compared with the operation of tactile sensor modeled by computer software. The appearance of an overshoot coincides exactly with the tumor location in the stress graphs and it points out that the experimental results verify the numerical ones reasonably well.

9.5 Graphical Rendering of Localized Lumps for MIS Applications

In this section, a system for characterizing embedded lumps and rendering them graphically is introduced. In the proposed system, using an MIS multifunctional tactile sensor,[21] the masses within the tissue are detected and located; then the features are extracted and are visually displayed. The utilized sensor unit is a comparatively

FIGURE 9.12 Experimental setup: (A) MOTOMAN robot, (B) Power supply, (C) Monitor or tactile display, (D) Electronic board, (E) Serial port, (F) Robot grippers, (G) Tactile probe, and (H) Phantom of tissue including tumor. (See also color insert.)

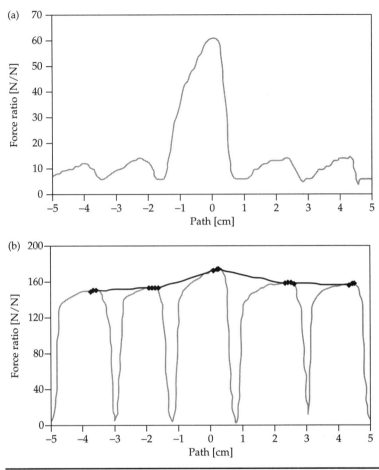

Figure 9.13 Experimental result for a model with (a) small h and (b) large h. [From Hosseini (2008),[4] courtesy of Science Publications.]

simple piezoelectric Polyvinylidene Fluoride (PVDF) based tactile sensor which is integrated with a new approach for the graphical display of lumps embedded in a soft object.[22] The displayed images are readily recognizable by the surgeon, and there is no need for extra hardware. When an array of these sensors is placed in one jaw of the endoscopic grasper, the location of the lump along the grasper can be detected and displayed. While using sensor arrays in both the upper and lower jaws of the grasper, it is possible to find and graphically represent the depth of the lump in the grasped object as well. Therefore, using the proposed system, surgeons can detect the presence or absence of the lump and obtain useful information on its size and location by simply grasping the target organ with the smart endoscopic grasper.

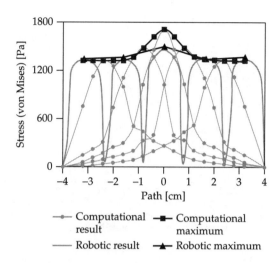

FIGURE 9.14 Operational comparison between computational and robotic results. (See also color insert.)

— Computational result
— Computational maximum
— Robotic result
— Robotic maximum

System Design

The proposed system used for this matter consists of a smart endoscopic grasper equipped with an array of tactile sensors, a data acquisition interface (DAQ), and the necessary signal processing algorithms that process the tactile information provided by the sensors and display the information visually. The complete system is schematically shown in Fig. 9.15. As illustrated in this figure, when a surgeon uses the smart endoscopic grasper to grasp a tissue, the sensor array measures the contact force distribution across each sensing element, as well as the total applied force. The electrical outputs of the piezoelectric sensing elements are then conditioned and transmitted to the data acquisition system. Using the data acquisition card, the signals are amplified, filtered, digitized, and processed by a computer. A computer code was developed in LabView software environment for signal conditioning such as filtering out line noise. In addition, a rendering algorithm, also developed in LabView software, was applied to map the extracted signal's features to a gray scale image. Using the constructed images, the surgeon discerns, not only the presence or absence of a lump, but also the approximate size and location of any lump that is detected.

Sensor Structure

The structure of the sensor integrated with an MIS grasper, as shown in Fig. 9.16, is corrugated to ensure that the tissue is grasped firmly. Figure 9.16 shows the proposed grasper, in which only the lower jaw is equipped with an array of tactile sensors. The number of sensors, their length, width, and thickness, as well as the space between them, could be optimized for any particular application. In this method, in order to replicate the human figure spatial resolution, the sensor array was made of seven equally spaced piezoelectric PVDF based

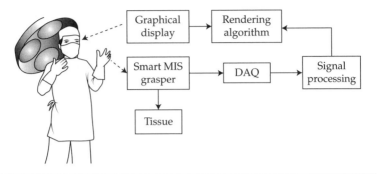

FIGURE 9.15 Components of the proposed system. (Sokhanvar, Ramezanifard, Dargahi, Packirisamy, "Graphical Rendering of Localized Lumps for MIS Applications," *Journal of Medical Devices*.[22] Copyright © 2007, Reproduced with permission of ASME International.)

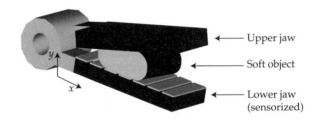

FIGURE 9.16 A view of the grasper with one active jaw equipped with an array of the seven sensing elements. (Sokhanvar, Ramezanifard, Dargahi, Packirisamy, "Graphical Rendering of Localized Lumps for MIS Applications," *Journal of Medical Devices*.[22] Copyright © 2007, Reproduced with permission of ASME International.) (See also color insert.)

sensing elements. The spatial resolution for a human finger using the two-point discrimination threshold (TPDT) is reported to be about 2 mm. Therefore, 1.5 mm wide sensing elements positioned 0.5 mm apart were considered for this proposition. Clearly, a finer array would yield better spatial resolution than that of the human finger.

Rendering Algorithm

Two sensor configurations on the grasper are examined. The grasper structure shown in Fig. 9.16, with one jaw equipped with sensors, is capable of locating the lumps in one dimension (x-axis); the grasper with two jaws equipped with sensors can characterize the lumps in two dimensions (x- and y-axes where the location of the lump in the y-direction can be considered as lump depth). In both designs, when there is no extraneous object in the soft tissue, depending on the grasper geometry and design, all sensing elements show either an equal output voltage or exhibit a regular pattern that is assumed as the background frame. The presence of the lump causes an uneven voltage distribution through the sensing elements. The deduction of the background frame from the total response gives the net effect of the lump and will increase the sensitivity. In addition to detecting lumps, the softness of the tissue in the sections with no embedded masses can also be measured.

The outputs of the sensing elements depend on several factors including: the ratio of the Young's modulus of the lump (E_L) to that of the tissue (E_T); the size and depth of the lump; and the magnitude of the applied load. Extracting all features of a lump using the minimal sensors detailed in this study is a difficult task, as some combinations of lump stiffness, size, depth, and the applied force create similar output patterns. This complexity is also reported in other references.[23] However, there are some constraints that can be used to reduce the number of variables, or at least to control their range. For instance, some analyses[24] show that, for $(E_L/E_T) > 10$, the variation of this ratio has only a negligible effect on the output. Fortunately, an absolute majority of the reported stiffness for the tumors are greater than this ratio.[23] Therefore, in practice, the output response is not greatly influenced by the variation of the Young's modulus of the lump. Another influential factor is the magnitude of the applied load. The contact force between the grasper and the tissue depends on the load exerted by grasper jaws on the tissue. Therefore, in addition to the pressure distribution, it is necessary to measure the total applied load. The applied load can be measured in different ways. For instance, a strain gauge attached to the jaw can provide data on the magnitude of the applied load. Another approach to measure the applied load was presented in other experimental work,[21] in which an extra PVDF film was used at the supports of each sensing element. In the experiments conducted in this method, the load was measured using a reference load cell. Furthermore, to reduce the number of contributing parameters throughout this approach, the force was kept constant. The other remaining factors are the size of the tumor and its location in both the x- and y-directions. Since the majority of masses can be approximated as spherical objects, the number of parameters needed to characterize the size of the sensor can be reduced to one value, such as the lump radius. The first design, as seen in Fig. 9.16, overlooks the depth of the lump and locates the lump merely in the x-direction. Instead, using two sets of arrays of sensing elements in the second design, it is possible to determine the depth of the lump as well.

1. Graphical representation of localized lumps in one dimension: As shown in Fig. 9.16, in the first design, the lower jaw is equipped with an array of sensors; hence the upper jaw only applies compressive load to the object containing lumps. To graphically represent the location of the lump, an image with seven vertical parallel bands corresponding to the seven sensing elements was initially considered, as in Fig. 9.17b. The intensity of each band was considered to be proportional to the output of the corresponding sensing element. The voltage distribution along the sensor array can be considered as a vector $\{V\}_{1\times7}$ that is related to the intensity vector $\{I\}_{1\times7}$ by Eq. (9.1):

FIGURE 9.17
Locating the lump in one direction and its graphical rendering.
(a) Sensor configuration details, (b) image prior to interpolation, and (c) image after interpolation. (Sokhanvar, Ramezanifard, Dargahi, Packirisamy, "Graphical Rendering of Localized Lumps for MIS Applications," *Journal of Medical Devices*.[22] Copyright © 2007, Reproduced with permission of ASME International.)

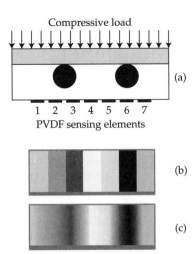

$$\begin{cases} I_i = \dfrac{V_i}{\alpha}(K-1), & V_i \le \alpha \\ I_i = K-1, & V_i \le \alpha \end{cases} \quad \text{for } i = 1, \ldots, 7 \quad (9.1)$$

where α is the normalizing factor that determines the working range (very soft, soft, medium, etc.), and K is the number of gray scales that are used in construction of a graphical image (here $K = 256$). It is seen from Eq. (9.1) that for a given α, when $V_i \le a$, the scaling factor α maps the input voltage domain into interval [0, 1]. Then this value, using the $(K-1)$ factor, would be mapped into the corresponding gray level, between 0 and 255. Once $V_i > \alpha$, all the values of V_i would be mapped to the maximum intensity (i.e., $I_i = 255$). For instance, Fig. 9.17b shows the graphical display for the case where two lumps were detected in the grasped tissue. In this case, one of the lumps had been positioned above sensing element 6 and the other one had been placed above and between sensing elements 2 and 3 (see Fig. 9.17a for configuration details). However, due to the limited number of the sensing elements, the quality of the image shown in Fig. 9.17b was not satisfactory. Therefore, by using an interpolation technique, the quality of the image was enhanced in Fig. 9.17c. Prior to applying the interpolation method, the number of elements had to be increased from 7 to any desired number (N). For this matter, (N–7) extra elements were required. Therefore, (N–7)/6 elements were inserted between each of the two original elements. The resulting $1 \times N$ vector $\{G\}$, has the following form:

$$\{G\} = \{G_1, G_2, \ldots, G_{N-1}, G_N\}$$

in which,

$$G_1 = V_1, \quad G_{\underline{N+5}\atop 6} = V_2, \quad G_{\underline{2N+4}\atop 6} = V_3, \quad G_{\underline{3N+3}\atop 6} = V_4,$$

$$G_{\underline{4N+2}\atop 6} = V_5, \quad G_{\underline{5N+1}\atop 6} = V_6, \quad G_N = V_7$$

The intensity values assigned to the inserted elements were calculated using linear interpolation relationship expressed in Eq. (9.2).

$$G_i = V_j + \left\{ i - 1 - (j-1)\left(\frac{N+5}{6}\right) \right\} \times \frac{V_{j+1} - V_j}{(N-1)/6}, \text{ for}$$

$$1 + (j-1)\frac{N-1}{6} < i < 1 + j\left(\frac{N-1}{6}\right) \tag{9.2}$$

where j ($\leq j \leq 6$) *and* i ($1 \leq i \leq N$) are indices associated with the original vector {V} and the augmented vector {G}, respectively. The numerical example for $N = 60$, is illustrated in Fig. 9.17c.

2. Graphical representation of localized lumps in two dimensions: Figure 9.18 illustrates the second type of grasper, in which both the upper and lower jaws are equipped with an array of sensors. Using this grasper, it is possible to locate lumps in two directions, along the jaw (x-axis), as well as its depth (y-axis).

The steps used for the construction of two-dimensional tactile images are demonstrated in a flowchart presented in Fig. 9.19.

For better clarification of the algorithms used in this approach, consider the case illustrated in Fig. 9.20a. This figure demonstrates a grasped tissue containing a lump that is aligned with the sensing elements 2_U and 2_L, where the subscripts *U* and *L* refer to the Upper and Lower sensing arrays, respectively. The distance of the lump from the upper and lower sensing elements are shown by labels *a* and *b*, respectively.

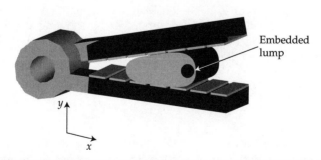

FIGURE 9.18 The second design of the grasper, where both upper and lower jaws are equipped with the sensing elements. (Sokhanvar, Ramezanifard, Dargahi, Packirisamy, "Graphical Rendering of Localized Lumps for MIS Applications," *Journal of Medical Devices*.[22] Copyright © 2007, Reproduced with permission of ASME International.)

Chapter 9

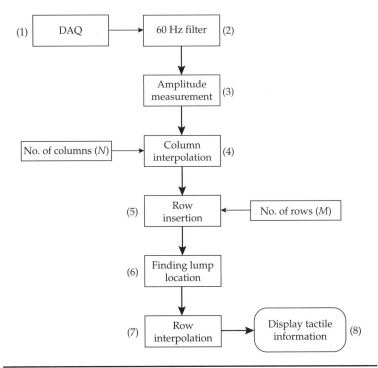

FIGURE 9.19 The algorithm flowchart implemented in LabView software used for graphical rendering. (Sokhanvar, Ramezanifard, Dargahi, Packirisamy, "Graphical Rendering of Localized Lumps for MIS Applications," *Journal of Medical Devices*.[22] Copyright © 2007, Reproduced with permission of ASME International.)

Fig. 9.20b shows the two-dimensional intensity graph, which was built using a one-dimensional algorithm as explained in the first design. This graph consists of two rows of color bands corresponding with the two sensor arrays, one on the top and the other at the bottom. Therefore, this graph can be considered as a matrix with 2 rows (color bands) and 7 columns (sensors), thereby showing 2 × 7 cells. The corresponding matrix, where each element represents voltage amplitude, has the following form:

$$[V] = \begin{bmatrix} V_{U1} & V_{U2} & V_{U3} & V_{U4} & V_{U5} & V_{U6} & V_{U7} \\ V_{U1} & V_{U2} & V_{U3} & V_{U4} & V_{U5} & V_{U6} & V_{U7} \end{bmatrix}$$

As it can be seen, Fig. 9.20b cannot clearly display valuable information about the lump location. To show the precise location of the lump, the dimensions of the matrix, and, consequently, the number of matrix elements, were increased. The graphical enhancement in the x-direction was explained in first design; thus, the row operations (in the y-direction) are emphasized in this section.

Tactile Sensing in Tumor Detection 161

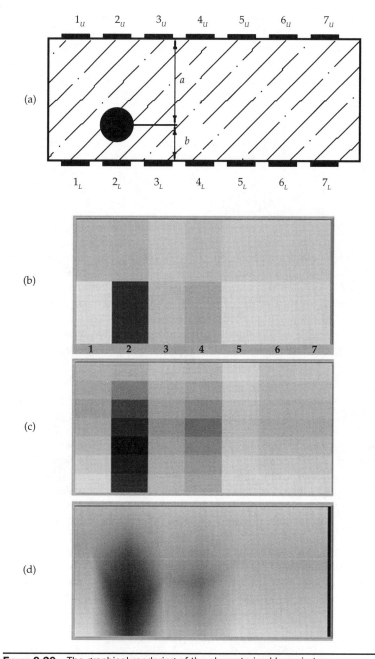

FIGURE 9.20 The graphical rendering of the characterized lump in two dimensions. (a) A lump located in a soft material with the upper and lower sensor arrays, (b) two-dimensional intensity graph associated with the sensor array outputs, (c) a 7 × 7 matrix showing the location of the lump, and (d) a 60 × 100 matrix that gives better information on the location and size of the lump. (Sokhanvar, Ramezanifard, Dargahi, Packirisamy, "Graphical Rendering of Localized Lumps for MIS Applications," *Journal of Medical Devices*.[22] Copyright © 2007, Reproduced with permission of ASME International.)

As shown in the flowchart of Fig. 9.19, Step 5, the number of rows was increased to M by inserting (M–2) rows of zeros between the first and second rows of matrix [V] which led to an M × 7 matrix. Furthermore, using the technique explained in first design, the number of columns was also increased to N. The resulting M × N matrix $[G_0]$ would be in the form of:

$$[G_0] = \begin{bmatrix} G_{U1} & G_{U2} & \cdots & G_{U(N-1)} & G_{UN} \\ 0 & 0 & \cdots & 0 & 0 \\ \vdots & \vdots & & \vdots & \vdots \\ 0 & 0 & \cdots & 0 & 0 \\ G_{L1} & G_{L2} & \cdots & G_{L(N-1)} & G_{LN} \end{bmatrix} \Bigg\} M \text{ Rows}$$

$$\underbrace{\phantom{G_{U1} \quad G_{U2} \quad \cdots \quad G_{U(N-1)} \quad G_{UN}}}_{N \text{ Columns}}$$

For the graphical representation of the lump, two parameters had to be determined; these were the location of the center of lump in each column and its corresponding intensity value. In order to designate the vertical location of the center of the lump in each column, as seen in Fig. 9.19, Step 6, a relationship between the thickness of the tissue and matrix rows $[G_0]$ was used. If a lump is located in the tissue at a distance of the upper sensor array, it will be mapped into row r, where r can be found from Eq. (9.3):

$$\frac{r}{M} = \frac{a}{a+b} = \frac{G_U}{G_L + G_U} \qquad (9.3)$$

In the above equation, $(a + b)$ is equal to the tissue thickness and was considered to be proportional to the number of rows (M). Regardless of the existence of a lump, Eq. (9.3) was applied to all columns, as seen in Fig. 9.21. If a lump exists in a column, then a and b are the distances of the center of the lump from the upper and lower sensor arrays, respectively. For the columns with no lump, the associated sensor outputs are equal and $G_U = G_L$, and thus $r = M/2$. These cells are indicated in Fig. 9.21 with gray color. In other words, the algorithm assigns a nonzero value to the middle row of the columns with no lump. Although this value is not significant, it can be considered as a shortcoming of the algorithm.

In order to determine the intensity values of these locations in each column, the following relation was used:

$$G_{rj} = G_{ui} + G_L$$

where index G_{rj} specifies the intensity value of the cell located in row r and column j, showing the center of the lump in that column. The result of this operation is matrix $[G_1]$, in which the centers of the detected lumps are specified.

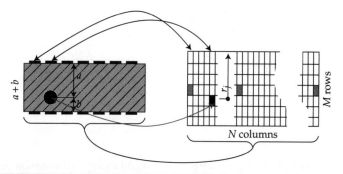

FIGURE 9.21 The relationship between grasped object and intensity matrix. (Sokhanvar, Ramezanifard, Dargahi, Packirisamy, "Graphical Rendering of Localized Lumps for MIS Applications," *Journal of Medical Devices.*[22] Copyright © 2007. Reproduced with permission of ASME International.)

$$[G_1] = \begin{bmatrix} G_{U1} & G_{U2} & \cdots & G_{UN} \\ 0 & 0 & \cdots & 0 \\ \vdots & \vdots & & \vdots \\ 0 & 0 & \cdots & 0 \\ G_{r_1 1} & G_{r_1 2} & \cdots & G_{r_1 N} \\ 0 & 0 & \cdots & 0 \\ \vdots & \vdots & & \vdots \\ 0 & 0 & \cdots & 0 \\ G_{L1} & G_{L2} & \cdots & G_{LN} \end{bmatrix} \} M \text{ Rows}$$

$$\underbrace{}_{M \text{ Columns}}$$

It should be noted that in the case of multiple lumps, the center of each lump will be mapped to a row that corresponds to the lump's original depth in the tissue. Therefore, for instance, $G_{r_1 1}$ and $G_{r_2 2}$ are not necessarily in the same row.

As shown in Fig. 9.19, Step 7, the next step is to implement a row interpolation procedure. At this step, three values are known in each column: G_{Ui}, G_{r_i}, and G_{Li}. Using these values and linear interpolation, a new intensity distribution can be assigned to all the zeros. The final intensity matrix [H] can be represented as:

$$[H] = \begin{bmatrix} H_{11} & H_{12} & \cdots & H_{1N} \\ H_{21} & H_{22} & \cdots & H_{2N} \\ \vdots & \vdots & \vdots & \vdots \\ H_{(M-1)1} & H_{(M-1)2} & \cdots & H_{(M-1)N} \\ H_{M1} & H_{M2} & \cdots & H_{MN} \end{bmatrix}$$

where the intensity of each cell is calculated from the following equation:

$$\begin{cases} H_{ij} = G_{Uj} + (i-1) \times \dfrac{G_{r_j j} - G_{Uj}}{r_j - 1}, & 1 \le i \le r_j, \ 1 \le j \le N \\ H_{ij} = G_{r_j j} + (i - r_j) \times \dfrac{G_{r_j j} - G_{Uj}}{r_j - 1}, & r_j \le i \le M, \ 1 \le j \le N \end{cases}$$

Figure 9.20c shows the lump position and its approximate size after implementing the mentioned algorithm when $M = N = 7$. Evidently, increasing the number of cells in both directions will enhance the quality of image. Figure 9.20d, for instance, is the constructed graphical image based on the same sensor's output and enhancement of the associated matrix to $M = 60$ and $N = 100$.

Experiments

An experimental setup was used to generate tactile information by applying known loads through the fabricated graspers to a soft object which contained a lump. The graspers were positioned under a probe which was equipped with a reference load cell, while the soft object and lump were compressed between two jaws. The photographs of both prototyped graspers, with one and two active jaws, are shown in Fig. 9.22a and Fig. 9.22b, respectively. Since the PVDF-based sensing elements were prepared manually, there was an initial discrepancy between the output voltages for equal loads. To compensate for this disparity, a controllable coefficient for each sensing element was defined. Then, using homogenous elastomeric materials without any inclusion, the outputs of the sensors were identically adjusted. The output voltages of the sensing elements in both designs were processed and graphically displayed according to the explained algorithm.

The soft elastomeric material with a known Young's modulus was used as the bulk soft object, and metallic balls simulating the lumps with different sizes (3.9, 6.3 and 7 mm) were inserted into the hollow spaces carved out of the elastomeric bulk. To change the depth of the lumps, several layers of the elastomeric material were cut into the same dimensions, but with different thicknesses. The lumps were placed in one of the layers, so that the other elastomeric layers were used as spacers to increase or decrease the distance of lump layer from the top and bottom surfaces. A dynamic load was applied by the shaker that was driven by a power amplifier and a signal generator, as shown in Fig. 9.23. The outputs of the sensors were fed into the connector box through electronic buffers. The piezoelectric PVDF can be considered as a voltage source with very high output impedance. Since the DAQ needs the input impedance to be less than 100 kΩ, a buffer was necessary to achieve impedance matching. Data was

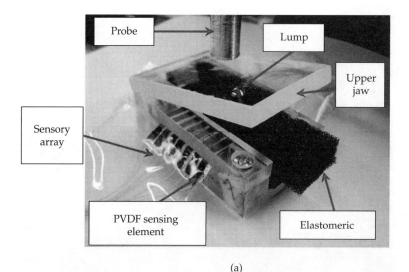

FIGURE 9.22 Photographs of the sensors under the test. (a) The sensor with one active jaw used for construction of one-dimension graphical images. (b) The sensor with two active jaws used for two-dimension graphical rendering of detected lumps. (Sokhanvar, Ramezanifard, Dargahi, Packirisamy, "Graphical Rendering of Localized Lumps for MIS Applications," *Journal of Medical Devices*.[22] Copyright © 2007, Reproduced with permission of ASME International.) (See also color insert.)

transferred to the computer using the DAQ (NI PCI-6225). The DAQ main amplifier was used in RSE (reference single ended) mode. A low pass filter with a cutoff frequency of 40 Hz was used to remove

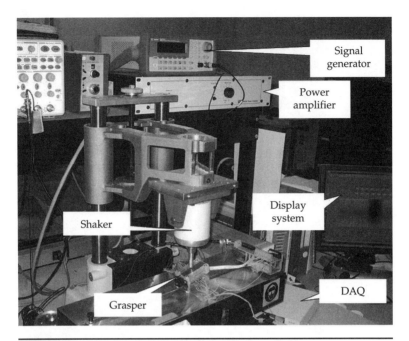

FIGURE 9.23 Photograph of the experimental setup. (Sokhanvar, Ramezanifard, Dargahi, Packirisamy, "Graphical Rendering of Localized Lumps for MIS Applications," *Journal of Medical Devices*.[22] Copyright © 2007, Reproduced with permission of ASME International.)

the 60 Hz line noise. As explained, the processing algorithm in the two designs was developed in LabView software environment for graphical demonstration of the forces sensed by the sensor elements.

Results

Two finite element models were developed to analyze the various types of graspers. The results of the finite element analysis, as well as the graphical representations of tactile information obtained from experimental cases, are shown in Fig. 9.24 and Fig. 9.25 for one- and two-dimensional procedures, respectively. Each row in Fig. 9.24 shows a scenario in which multiple lumps with different sizes were inserted into the elastomeric bulk. The left column in this figure illustrates the geometrical information about the locations and sizes of the lumps that were placed in the soft object. The middle column in Fig. 9.24 is a one-dimension graphical representation of the sensor's outputs obtained from the experiments. The right column represents the normalized voltage response of the sensing elements obtained from the finite element analysis.

In the graphical representation in Fig. 9.24*a* (middle column), the dark column 2 has the highest intensity, showing that the lump is located

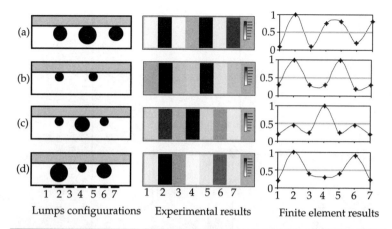

FIGURE 9.24 The experimental and analytical results of four case studies. (Sokhanvar, Ramezanifard, Dargahi, Packirisamy, "Graphical Rendering of Localized Lumps for MIS Applications," *Journal of Medical Devices*.[22] Copyright © 2007, Reproduced with permission of ASME International.)

above this sensing element. This can be compared with the intensity of sensing elements 4 and 5 that share a lump. For the latter elements, the maximum contact stress value occurs in a place between sensing elements 4 and 5. Therefore, each sensing element senses part of the load and shows lower amplitude, in comparison with sensing element 2.

These two elements also provide information about the size of the lump. If the middle lump was large enough to cover both sensing elements, the result would be two completely dark bands. Therefore, from the shown gray levels, the approximate size of the middle lump can be deduced. The difference observed between the outputs of sensing elements 2 and 7 can be attributed to the edge effect on the latter element. The second case, Fig. 9.24*b*, shows two identical lumps embedded above sensing elements 2 and 5. A similar output voltage and intensity can be seen in the graphical representation, as well as the finite element analysis. In the third case, in Fig. 9.24*c*, a larger lump is placed between two smaller lumps. It is shown that the system is capable of detecting all three masses. However, the darker band associated with sensing element 4 gives information on the relative size of the middle lump with respect to the other lumps. In the last case, in Fig. 9.24*d*, a small lump had been positioned between two larger lumps. As the results show, the sensor was not able to detect the smaller mass. A closer examination of finite element stress distribution shows that two larger lumps created a stress profile between themselves in such a way that the effect of the small mass was suppressed. This figure demonstrates that, for multiple lumps with different sizes and locations, more than one attempt in different orientations might be needed to obtain an accurate result.

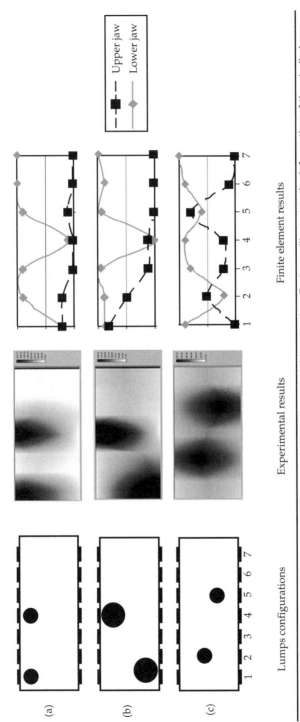

FIGURE 9.25 The experimental and analytical results for two-dimensional localization. Each row illustrates the information of the studied case. In the right column, the dashed line represents the output voltages of the lower array of the sensors, while the solid line is associated with the upper jaw. (Sokhanvar, Ramezanifard, Dargahi, Packirisamy, "Graphical Rendering of Localized Lumps for MIS Applications," *Journal of Medical Devices*.[22] Copyright © 2007, Reproduced with permission of ASME International.)

Figure 9.25 shows the results of four case studies in which both jaws of the graspers were equipped with arrays of sensing elements. Figure 9.25a demonstrates a case in which two identical lumps were positioned close to sensing elements 1 and 4 of upper jaw (1U and 4U, respectively). The corresponding graphical image shows clearly the location of the lumps. In addition, the gray level of the image gives some information about the size of the lumps.

In the next configuration, in Fig. 9.25b, lumps are positioned apart in such a way that one lump is put under sensing element 4U, while the second lump is placed above sensing element 1L (slightly overlapped with sensing element 2L). The graphical image constructed based on the experimental sensor's output is shown in the middle column of Fig. 9.24b, in which the location (in x- and y-directions) is clearly extractable. The gray levels in this case can be compared with those of Fig. 9.25c, in which the lumps were positioned far from the sensing elements. Again the position and size of lumps can be clearly perceived from the experimental data shown in the middle column. The finite element results shown in the right column are consistent with the experimental data in all three cases. However, implementation of this algorithm may produce gray areas in the middle of images in the absence of any lump. To rectify this problem, an enhanced algorithm is under development.

Graphical rendering of localized objects is a feasible technique with great potential for use in MIS. Using this method, a part of the tactile information which is lost when palpation is not done, can be restored. This capability is useful, not only for MIS, but also for MIS robotic surgery and, in general, for robotic surgery. Other anatomical features such as beating arteries can potentially be detected and graphically rendered.

References

1. J. Dargahi and S. Najarian, "Human Tactile Perception as a Standard for Artificial Tactile Sensing—A Review," *International Journal of Medical Robotics and Computer Assisted Surgery*, vol. 1, no. 13, 2004, pp. 23–35.
2. J. Dargahi and S. Najarian, "Advances in Tactile Sensors Design/Manufacturing and Its Impact on Robotics Application—A Review," *Industrial Robot*, vol. 32, no. 3, 2005, pp. 268–281.
3. S. M. Hosseini, S. Najarian, S. Motaghinasab, and S. Torabi, "Experimental and Numerical Verification of Artificial Tactile Sensing Approach for Predicting Tumor Existence in Virtual Soft Tissue," *Proceedings of the 15th Annual-International Conference of Mechanical Engineering*, Amirkabir University of Technology, Tehran, Iran, 2007.
4. S. M. Hosseini, S. Najarian, and S. Motaghinasab, "Analysis of the Effects of a Tumour in the Biological Tissue Using Artificial Tactile Sensing Modeling," *Amirkabir Journal of Science and Technology* (in press), 2006.
5. S. Najarian, J. Dargahi, and V. Mirjalili, "Detecting Embedded Objects Using Haptics with Applications in Artificial Palpation of Tumours," *Sensors & Materials*, vol. 18, no. 4, 2006, pp. 215–229.
6. S. M. Hosseini, S. Najarian, S. Motaghinasab, and J. Dargahi, "Detection of Tumors Using Computational Tactile Sensing Approach," *International*

Journal of Medical Robotics and Computer Assisted Surgery, vol. 2, no. 4, 2006, pp. 333–340.
7. Z. Huo, M. L. Giger, and C. J. Vyborny, "Automated Computerized Classification of Malignant and Benign Masses on Digitized Mammograms," *Academic Radiology*, vol. 5, no. 3, 1998, pp. 155–168.
8. Z. Huo, M. L. Giger, C. J. Vyborny, D. E. Wolverton, and C. E. Metz, "Computerized Classification of Benign and Malignant Masses on Digitized Mammograms: A Study of Robustness," *Academic Radiology*, vol. 7, no. 12, 2000, pp. 1077–1084.
9. B. Sahiner, H. P. Chan, N. Petrick, M. A. Helvie, and M. M. Goodsitt, "Computerized Characterization of Masses on Mammograms: The Rubber Band Straightening Transform and Texture Analysis," *Medical Physics*, vol. 25, no. 4, 1998, pp. 516–526.
10. A. Retico, P. Delogu, M. E. Fantacci, and Kasae P, "An Automatic System to Discriminate Malignant from Benign Massive Lesions on Mammograms," *Nuclear Instruments & Methods in Physics Research*, vol. 569, 2006, pp. 596–600.
11. B. Verma and P. Zhang, "A Novel Neural-Genetic Algorithm to Find the Most Significant Combination of Features in Digital Mammograms," *Applied Soft Computing*, vol. 7, no. 2, 2006, pp. 612–625.
12. C. S. Tong, "Blur Detection Using a Neural Network," *Proc SPIE*, 1995, pp. 348–358.
13. G. De Nicolao, G. Sparacino, and C. Cobelli, "Nonparametric Input Estimation in Physiological System: Problems, Methods and Case Studies," *Automatica*, vol. 33, no. 5, 1997, pp. 851–870.
14. J. T. Ives, R. F. Gesteland, and J. R. Stockham, "An Automated Film Reader for DNA Sequencing Based on Homomorphic Deconvolution," *IEEE Transactions on Biomedical Engineering*, vol. 41, 1994, pp. 509–519.
15. G. De Nicolao and G. Ferrari-Trecate, "Regularization Networks for Inverse Problems: A State–Space Approach", *Automatica*, vol. 39, 2003, pp. 669–676.
16. M. H. Lee and H.R. Nicholls, "Tactile Sensing for Mechatronics—A State-of-the-Art Survey," *Mechatronics*, vol. 9, no. 1, 1999, pp. 1–31.
17. M. E. H. Eltaib and J. R. Hewit, "Tactile Sensing Technology for Minimal Access Surgery—A Review," *Mechatronics*, vol. 13, no. 10, 2003, pp. 1163–1177.
18. S. Tez, Ö. Yolda, Y. Klç, H. Dizen, and M. Tez, "Artificial Neural Networks for Prediction of Lymph Node Status in Breast Cancer Patients," *Medical Hypotheses*, vol. 68, no. 4, 2006, pp. 922–923.
19. S. M. Hosseini, M. Amiri, S. Najarian, and J. Dargahi, "Application of Artificial Neural Networks for Estimation of Tumor Characteristics in Biological Tissues," *International Journal of Medical Robotics and Computer Assisted Surgery*, vol. 3, 2007, pp. 235–244.
20. S. M. Hosseini, S. Najarian, S. Motaghinasab, A. Tavakoli Golpaygani, and S. Torabi, "Prediction of Tumor Existence in the Virtual Soft Tissue by Using Tactile Tumor Detector," *American Journal of Applied Sciences*, vol. 5, no. 5, 2008, pp. 483–489.
21. S. Sokhanvar, M. Packirisamy, and J. Dargahi, "A Novel PVDF Based Softness and Pulse Sensor for Minimally Invasive Surgery," *The Third IEEE International Conference on Sensors*, Austria, 2004, pp. 24–27.
22. S. Sokhanvar, M. R. Ramezanifard, J. Dargahi, and M. Packirisamy, "Graphical Rendering of Localized Lumps for MIS Applications," *Journal of Medical Devices*, vol. 1, no. 3, 2007, pp. 217–224.
23. P. S. Wellman and R. D. Howe, "Extracting Features from Tactile Maps," *Proceedings of the Second International Conference on Medical Image Computing and Computer-Assisted Intervention*, 1999, pp. 1133–1142.
24. S. Sokhanvar, J. Dargahi, and M. Packirisamy, "Nonlinear Modeling and Testing of Soft Tissue Embedded Lump for MIS Applications," submitted to the *International Journal of Medical Robotics and Computer Assisted Surgery*.

CHAPTER 10
Determination of Mechanical Properties of Biological Tissues Including Stiffness and Hardness

10.1 Introduction

Different methods are available for investigating the properties of soft tissues. Gao et al. reviewed some of the previous work done, which is mostly in the related field of biomechanics and measurement techniques.[1] First, they considered elastography as an important method. It is a merger of several related fields of study. These fields are: tissue biomechanics, tissue contrast differences, tissue motion by using imaging systems (X-ray, ultrasound, and MRI), and vibrating targets by using coherent radiation (laser, sonar, and ultrasound). Additionally, in recent years, atomic force microscopy (AFM) has been successfully applied to local elasticity measurements.[2] This technique is especially used in the field of biology. In this method, a cantilever is used which causes some problems in measurements and sample preparation. Another disadvantage of AFM is its high cost, resulting in a limited industrial and clinical use of AFM.[2] Tactile sensing technology can also be employed for the measurement of soft tissue elasticity. This technology tries to imitate the sense of touch. In the fields of minimal invasive surgery (MIS) and robotic surgery, this sense is of crucial importance.

In determining the physical properties of living tissues, stiffness plays a great role. As an example, a surgeon can determine if a cartilage or myocardial tissue is healthy by touching and determining its stiffness.

Determining the Stiffness of Cartilage

The three major constituents of an articular cartilage are chondrocytes, extracellular matrix, and interstitial fluid. In the extracellular matrix, proteoglycans primarily determine the compressive stiffness of the cartilage. The tensile property of the cartilage is determined by collagen fibers. A decrease in the proteoglycans content and an increase in the amount of fluid might initially alter the cartilage structure. This results in the cartilage becoming softened. Following this, fissures and cracks may appear in the cartilage once the proteoglycans content decreases. The final outcome is that the cartilage deteriorates and degenerates.[3]

Determining the stiffness of a cartilage can help immensely for detecting the first signs of cartilage degeneration. Many devices and equipment have been developed in order to measure the stiffness of cartilage samples that are obtained *in vitro* or by autopsy (necropsy). Despite this, few studies have been conducted on the stiffness of a cartilage in living humans.

Tactile Sensor System

Stiffness of the cartilage was measured with a tactile sensor system (Biosensor system, AXIOM Co Ltd, Fukushima, Japan). It can quantitatively assess the degree of softness or hardness of materials through an ultrasonic tactile sensor. This system is washable, can be sterilized, and fulfills all electrical safety regulations for medical instruments. The method employed involves measuring the change in resonance frequency. For this to happen, a vibrating ceramic transducer touches the surface of an object.[3]

Each material has its own resonant frequency, and if a specific material touches an object when vibrating at this frequency, there will be a shift in frequency. The stiffness or hardness of the object is related to the difference between the frequencies under nontouching and touching conditions.[3] Hence, the stiffness can be estimated by monitoring the shift in frequency. This is possible because this frequency shift depends on the stiffness of the object. Generally, a typical tactile sensor system of this type is composed of a sensor probe, an amplifier, and a filter, as seen in Fig. 10.1.

The sensor probe is 11 cm long, 4 mm in diameter, and weighs 2.08 g. It is equipped with a small tip having a diameter of 3 mm. This tip is connected acoustically to a piezoelectric transducer made of lead zirconate–barium titanate ceramic. The resonance frequency of the ceramic is 68 kHz. It is reported that measurements, using a frequency counter device, were made 150 times per second.[3]

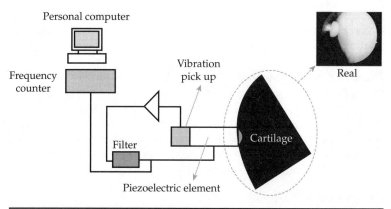

Figure 10.1 Tactile sensor structure. The change in resonance frequency when the sensor tip is attached to the tissue is a measure of the acoustic impedance of the tissue. (Reprinted from *Medical Engineering & Physics*, vol. 24, Uchio and et al. "Arthroscopic Assessment of Human Cartilage Stiffness of the Femoral Condyles and the Patella with a New Tactile Sensor," pp. 431–435, Copyright © 2002, with permission from Elsevier.) (See also color insert.)

It is of clinical important to evaluate the regional myocardial contractile function. For accurate measurement of myocardial stiffness *in situ*, a tactile sensor system has been developed.[3] It was found that the myocardial stiffness, or the tactile stiffness, measured by this sensor can be a very useful index for accurately quantifying regional myocardial function. Here, a coronary stenosis model was used to investigate regional myocardial tactile stiffness. This investigation was made under conditions of decreased coronary blood flow. Under these conditions, it was found that the regional myocardial tactile stiffness is a useful index sensitive enough to detect changes in regional contractile function.[4]

10.2 Experimental and Theoretical Analysis of a Novel Flexible Membrane Tactile Sensor

A new type of tactile sensor, which can detect both the contact-force and the stiffness of an object, has been recently proposed.[5] The major advantages in the proposed system are its simplicity and its robust design.

Sensed Objects

In tactile sensing, a force range between 0.1 N to 10 N is considered to have practical applications in medical devices. Also, a review of related literature shows that the variations of compliance and stiffness are quite large in different biological tissues. For instance, the Young's modulus of elasticity is about 0.11 MPa for the pig spleen, while it is about 4.0 MPa for the pig liver. Parallel instances also occur in human tissues.

Two-Dimensional Surface Texture Image Detection

Figure 10.2 shows an array of two elements of tactile sensors. When the tactile sensor array comes into contact with an object that has a bumpy surface, some of the mesa structure on the membrane pushes inwards; as a result, the system can detect the presence or absence of an object above each of the sensor elements. This enables us to obtain a two-dimensional surface texture image of the object.

Contact-Force Estimation

The structure of the estimated contact-force is also shown in Fig. 10.2. When the mesa of membrane comes into contact with an object, the normal force or uniform pressure from it causes inward deformation of the membrane. Therefore, by determining the displacement at the center of the membrane and according to the mechanical properties, the amount of normal force or the uniform pressure actuating on it is measurable.

Stiffness Detection

In this mode, the contacting mesa elements are pneumatically driven against the object, as seen in Fig. 10.3. The contact regions of the object are deformed according to the driving force of the mesa element and the stiffness of the object. Therefore, the stiffness of the object can be detected by measuring the relationship between the deflection of the membrane and its actuation force.

The operation of the tactile sensor has been analyzed theoretically and numerically, and the results are compared with the experimental results.

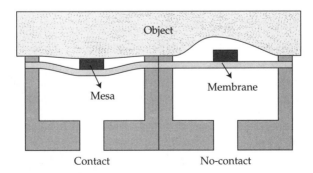

Figure 10.2 An array of two elements for detecting contact-force distribution and surface texture image. (Redrawn from Emamieh 2008,[5] courtesy of Science Publications.)

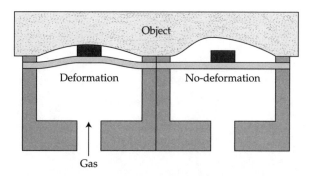

Figure 10.3 A schematic of the tactile sensor for detecting stiffness distribution. [Redrawn from Emamieh (2008),[5] courtesy of Science Publications.]

Device Specification

The device has a cylindrical shape which simplifies the problem and reduces the amount of calculation required. The radius of the membrane is 2 cm and is attached on a rigid cylinder which has a gas supply port and exhaust. The membrane thickness is 100 μm and the mesa radius is 0.5 cm with a thickness of about 150 μm. Table 10.1 shows the typical specifications of the sensor element.

Theoretical Analysis

The problem of large axi-symmetrical deformation of a circular membrane has practical significance. From the large deformation theory of a clamped single-layer circular membrane under the concentrated force, as seen in Fig. 10.4, the solution for out-of-plane deflections (OPD) can be expressed as Eq. (10.1):

$$\left(\frac{\Delta y}{t}\right)^3 = \left[1 - \left(\frac{1-3v}{4}\right)^{1/3}\right]^3 \frac{4R^2}{(1+v)\pi E h^4} F \qquad (10.1)$$

where Δy is out-of-plane deflection of the membrane, R is membrane radius, t is membrane thickness, v is Poisson's ratio, E is elastic modulus, and F is applied force at a central point.[6]

If $v = 1/3$, Eq. (10.2) can be yielded as below:

$$\left(\frac{\Delta y}{t}\right)^3 = \frac{3R^2}{\pi E h^4} F \qquad (10.2)$$

The tactile sensor has been theoretically analyzed for determining the stiffness of an object and assumed that the membrane and contacted object are elastic materials. Therefore, the membrane and the contacted object were modeled with two springs under a concentrated force.

Device (cylindrical)	2 cm (inner radius)
	3 cm (outer radius)
	5 cm (height)
Membrane	2 cm (radius)
	100 μm (thickness)
Mesa	0.5 cm (radius)
	150 μm (thickness)
v (Poisson's ratio)	0.33
E (Elastic modulus)	30 MPa

Source: (From Emamieh (2008),[5] courtesy of Science Publications).

TABLE 10.1 Typical Specifications of a Modeled Sensor Element.

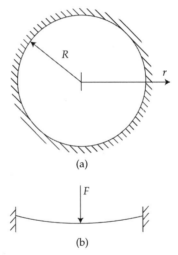

FIGURE 10.4 Theoretical model (a) front view (b) top view. [Redrawn from Emamieh (2008),[5] courtesy of Science Publications.]

Experimental Method

For research in biological and physiological areas, accurate measurement of tissue mechanical properties is required, although online monitoring is not strictly necessary. In most medical devices, we require an instrument that is able to provide the information in real-time, even if it is less accurate. This is especially true in diagnostic instruments, such as for the identification of abnormality in the mechanical properties of an organ or tissue, e.g., stiffness.

A single tactile sensor was fabricated according to the theoretical model specifications. The membrane material of ideal elastic (no hysteresis) should yield sufficiently in a repeatable manner, under low force.

Several polymers were studied to identify one that possessed suitable mechanical properties and was more compatible with biological tissues. Materials with a high hysteresis were rejected. The sensor body (substrate of membrane) was made of PVC because of its rigid functionality and low cost. Also, a particular kind of silicon rubber was chosen as the membrane, and different types of silicon rubber with different hardness were selected for the samples to test the tactile sensor.

Samples with great stiffness, less stiffness, and similar stiffness as the membrane were selected. Table 10.2 demonstrates some mechanical properties of membrane and samples.

Shapes of the samples were rectangular and were fabricated using molding technology. A single strain gauge was embedded in the membrane, exactly in the peripheral and radius directions of the membrane. It should be noted that the strain gauge is one of the most important tools in electrical measurement techniques applied to the measurement of mechanical quantities.

The embedded strain gauge (Tokyo Sokki Kenkyujo, 120 Ω, FLA-10-11) on the membrane is connected to a Wheatstone bridge circuit consisting of a single strain gauge (Quarter Bridge). The axi-symmetric shape of the sensor is useful for simplifying since it reduces the number of bonded strain gauges to only one. The Wheatstone bridge, which is connected to a stabilized DC power supply, and its amplifying circuit is shown in Fig. 10.5.

Force applied to the center of the membrane causes it to stretch; the embedded strain gauge then measures the resulting strain in the membrane. As a result of membrane deflection and resulting stress on the strain gauge, resistive changes take place and unbalance the Wheatstone bridge. This results in an output signal proportional to the stress value caused by membrane deflection. Since this output signal is small (in the order of 10^{-3} volts), an amplifying circuit is necessary to increase the signal to a level suitable for application and measurement.

In an experiment, a cylindrical probe driven by a robot hand (MOTOMAN, Yaskawa, Japan, 6 degrees of freedom) was used to apply a static force with a precise deflection on the membrane of the tactile sensor. In another experimental setup for determining

Material	Young's modulus (MPa)
Si rubber (I)	0.77
Si rubber (II)	3.75
Si rubber (III)	1.07
Polyurethane	548

Source: From Emamieh (2008),[5] courtesy of Science Publications.

TABLE 10.2 Specifications of Membrane and Samples.

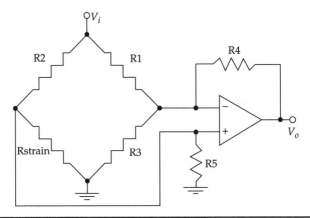

FIGURE 10.5 Wheatstone bridge circuit with amplifiers. [Redrawn from Emamieh (2008),[5] courtesy of Science Publications.]

the contact-force, static forces were applied to the sensor using steel weights, and the output from the strain gauge was recorded. Loads were applied incrementally to nearly 1 N and then unloaded.[5]

A compressor and other ancillary equipment, including a needle valve, pressure gauge, and pipe, were used for another experimental setup in order to estimate the stiffness of the fabricated samples.

Results

Figure 10.6 plots the measured output voltage of the strain gauge versus the applied force to the membrane, while Fig. 10.7 plots the measured output voltage of the strain gauge versus the deflection at the center of the membrane.

In addition, the changes of the membrane deflection in contact with different materials were investigated and the following results were obtained:

1. If the elasticity of an object is much smaller than the elasticity of the membrane, the sensor cannot sense the stiffness of the object and its variations.

2. If the elasticity of an object is too large compared to the membrane, the deflection of the membrane is very small and, with increasing stiffness of the object, the amount of deflection declines to zero. As a result, in this region, the membrane of the sensor cannot be deformed and the sensor cannot detect the stiffness or its changes.

3. If the mechanical properties of an object are similar to those of the membrane, the amount of deflection in the membrane is proportional to the stiffness of the object and this amount changes with the variations of stiffness. As a result, in

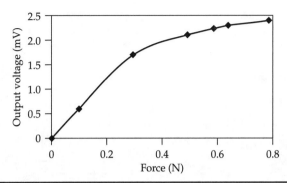

FIGURE 10.6 Output voltage versus the applied force. [From Emamieh (2008),[5] courtesy of Science Publications.]

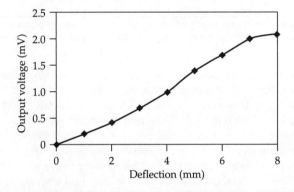

FIGURE 10.7 Output voltage versus the out-of-plane deflection. [From Emamieh (2008),[5] courtesy of Science Publications.]

this region, the sensor can detect the stiffness of the object according to the membrane deflection, against the amount of deflection in the same condition as when membrane has no contact with the object.

It was concluded that in order to detect a change in the stiffness of the touched object, the elasticity of the membrane should be almost similar to that of the touched object.

Figure 10.8 shows the measured output voltage of the electronic circuit of the sensor versus the gas pressure applied to the membrane with no contact and contact cases with different samples. The output voltage in all cases drastically increased with increasing gas pressure.

As shown in Fig. 10.8, the amount of deflection of the membrane in contact with the stiffer objects at unique pressure is less than that of the others. So, moving the sensor on an identified area at a constant

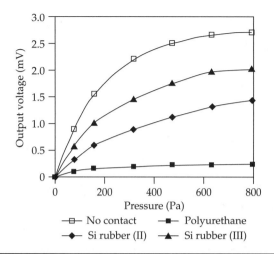

FIGURE 10.8 Variations of output voltage at different gas pressures. [From Emamieh (2008),[5] courtesy of Science Publications.]

pressure, the change in membrane deflection, and hence the output voltage of sensor, describes the changes of surface stiffness of that area.

Following this work, another modified tactile sensor was presented, as shown in Fig. 10.9. When the tactile sensor array comes into contact with an object that has a bumpy surface, some of the mesa structures on the membrane push inwards, causing a change in capacitance. Hence, the system can detect the presence or absence of an object above each of the elements; we are, therefore, able to obtain a two-dimensional surface texture image of the object.[7]

10.3 A Micromachined Active Tactile Sensor for Hardness Detection

Principle of the Tactile Sensor

The structure of a micromachined tactile sensor is shown in Fig. 10.10. This sensor can detect both the contact-force and the hardness of an object. The main components are: a diaphragm with a mesa (a flat elevated area) at the center, a piezo-resistive strain sensor at the periphery, and a chamber for pneumatic actuation. The device specifications were designed to detect human finger touch. Using micromachining technology, the fabrication process of the device was developed. The size of the sensor element is 11.0 mm × 15.0 mm × 0.4 mm.[8] Differences in hardness in the range of 10^3–10^5 N/m can be detected by the fabricated tactile sensor.[8]

As shown in Fig. 10.11, when the tactile sensor array touches an object that has a bumpy surface, some of the mesa structures on the

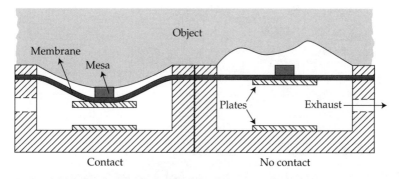

FIGURE 10.9 An array of two elements for detecting contact-force distribution and surface texture image. [From Tavakoli Golpaygani (2008),[7] courtesy of Science Publications.]

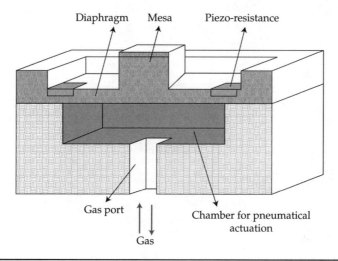

FIGURE 10.10 A micromachined tactile sensor. (Reprinted from *Sensors and Actuators*, vol. 114, Hasegawa et al. "A Micromachined Active Tactile Sensor for Hardness Detection," pp. 141–146, Copyright © 2004, with permission from Elsevier.)

diaphragms move to contact the object. This, in turn, results in the diaphragms being pushed inwards. Following this, the system can then detect the contact-force distribution. Additionally, some information on the 2D surface texture image of the object can be obtained.[8]

Figure 10.12 shows the second mode, in which the contacting mesa elements are pneumatically driven against the object. Due to the driving force of the mesa element and the hardness of the object, the contact regions of the object are deformed, as seen in Fig. 10.13. As a consequence, the hardness distribution of the object can be detected. This mode is based on measuring the relationship

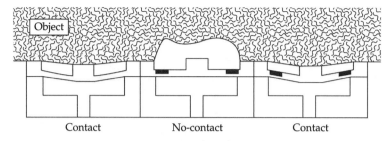

FIGURE 10.11 Arrayed tactile sensors for detecting force contact distribution and 2D surface image. (Reprinted from *Sensors and Actuators*, vol. 114, Hasegawa et al. "A Micromachined Active Tactile Sensor for Hardness Detection," pp. 141–146, Copyright © 2004, with permission from Elsevier.)

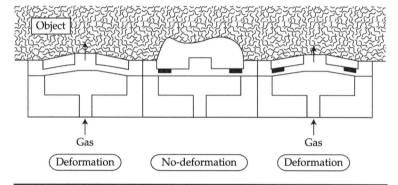

FIGURE 10.12 Arrayed tactile sensors for detecting hardness distribution. (Reprinted from *Sensors and Actuators*, vol. 114, Hasegawa et al. "A Micromachined Active Tactile Sensor for Hardness Detection," pp. 141–146, Copyright © 2004, with permission from Elsevier.)

between the actuation force of the elements and the deformation of the diaphragms.

10.4 Design and Fabrication of a New Tactile Probe for Measuring the Modulus of Elasticity of Soft Tissues

In this work, we discussed the design, fabrication, and testing of a new tactile probe called *Elastirob*,[9] which is used to measure the modulus of elasticity of biological soft tissues and soft materials. Elastirob determines the elasticity by drawing the stress-strain curve and then calculating its slope. The combination of the force sensing resistor (FSR), microcontroller, and stepper motor provides Elastirob with the ability to apply different rates of strain on testing specimens. Elastirob is accompanied by a tactile display called *TacPlay*.[9] This display is a custom designed, user friendly interface

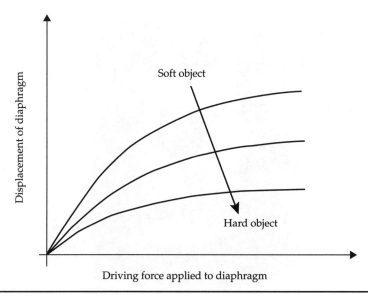

FIGURE 10.13 Displacement of diaphragm as a function of the hardness of the object. (Reprinted from *Sensors and Actuators*, vol. 114, Hasegawa et al. "A Micromachined Active Tactile Sensor for Hardness Detection," pp. 141–146, Copyright © 2004, with permission from Elsevier.)

and is able to evaluate the elasticity in each part of the stress-strain curve.

Introduction

Various properties of biological tissues have been the subject of interest in different medical applications; this is based on the fact that these properties contain a lot of useful information including age, gender, and whether or not the organ is healthy. Among these properties are Young's modulus (or stiffness), Poisson's ratio, and viscosity. Perhaps the most important parameter is the Young's modulus because of its dependence upon the composition of the tissue. Consequently, changes in soft tissue stiffness may be related to an abnormal pathological complication. Examples include breast, liver, and prostate cancers.

Description of the System

The tactile sensor system presented here consists of the following components: an FSR sensor; a stepper motor (including a converter which converts rotating motion to linear motion); an electronic board for processing data and controlling the system motion; and its unique TacPlay software, as seen in Fig. 10.14. This system is able to measure the modulus of elasticity of various kinds of soft tissues and biomaterials. Elastirob will greatly contribute to the field of tactile

184 Chapter 10

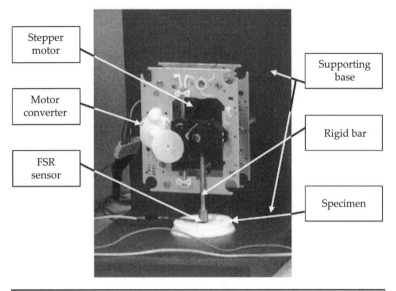

Figure 10.14 Tactile probe parts with a specimen in its place. [From Roham et al. (2007),[9] courtesy of Emerald Group Publishing.]

sensing technology, since all information is acquired just by simple physical contacts of the probe with testing specimens. Moreover, the specific Elastirob characteristics which make it an appropriate device to be developed for robotic surgery and telemedicine include: its low cost of construction; the capability of being miniaturized; the flexibility of programming; and changing the test adjustments because of using the combination of the FSR, microcontroller and stepper motor (see Fig. 10.15 for a typical output). In addition, TacPlay is very user friendly. It can both guide users through the measuring process with its appropriate prompts and can prevent possible mistakes. The capability of changing the step interval between every step of the stepper motor in TacPlay allows the operators to investigate the effect of strain rate and examine tissue response with regard to the modulus of elasticity of the soft tissues.[9]

10.5 Tactile Distinction of an Artery and a Tumor in a Soft Tissue by Finite Element Method

Despite the advantages of MIS and its growing popularity, it suffers from one major drawback; it decreases the sensory perception of the surgeon who may accidentally cut or incur damage to some of the tissues. This effect is more pronounced when the surgeon approaches a target tissue while moving across other healthy tissues. This could happen during grasping or manipulation of biological tissues such

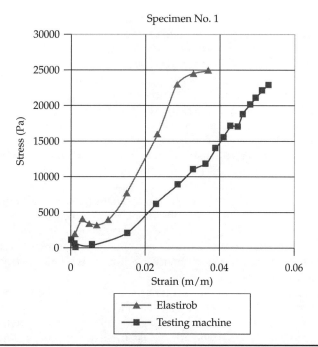

FIGURE 10.15 Comparison between the output of Elastirob and that of a commercial testing machine. [Redrawn from Roham et al. (2007),[9] courtesy of Emerald Group Publishing.]

as veins, arteries, or bones. Any inhibition on the surgeon's sensory abilities might lead to undesirable results. One of the main difficulties encountered in this area is the inability of detecting arteries embedded in tissues. Therefore, in laparoscopic surgeries such as bringing the gall bladder out of the body, the cutting location is either burnt for bloodshed prevention, or some grippers are placed in each side of the cutting location.

Although many studies have been performed on the detection of arteries in tissues and their stenosis, this subject is new. Nearly all of these studies have used imaging and ultrasound techniques. Besides their advantages, they also have some limitations including: vessels deep in the body are harder to see than superficial vessels; smaller vessels are more difficult to image and evaluate than larger vessels; any calcification that occurs as a result of atherosclerosis may obstruct the ultrasound beam; and sometimes ultrasound cannot differentiate between a blood vessel that is closed and the one which is almost closed, because the weak volume of blood flow produces a weak signal.[10] Therefore, a practical method which could eliminate these limitations during the surgical process is of great importance. Artificial tactile sensing is a new technique for detecting arteries in soft tissues by means of palpation.

Unlike tumor detection, only a very few number of studies could be found on detecting an artery and the location of its stenosis by tactile methods. In one of the studies related to tactile detection of arteries, a sensorized finger was constructed which was capable of detecting the pulse rate and waveform at wrist artery and sensing hard nodules in a mock breast.[11] It is important to note that in this method, numerical solutions were not performed. In another approach, a long 10 mm diameter probe was constructed with an array of tactile sensors set in the end of the probe.[12] This probe was pressed against the tissue of interest and the pressure distribution was converted into an electrical signal across contact area. This information was then processed and the presence of an artery in the tissue was determined. In this method, just as in all previous works carried out, numerical solutions were not employed. The last method on artery detection is related to construction of a tactile sensor that can track a vessel with various curves in artificial tissue of silicon type by a programmable robot.[13] Similar to previous works, numerical solutions were not performed when using this approach.

One recent study, and for the first time, investigated the detection of an artery and a tumor in a tissue and separation of these two categories from each other by using the finite element method.[10] Furthermore, a 25% stenotic artery was modeled in a tissue and the possibility of distinction of a healthy artery from a stenotic artery was investigated and presented in the form of a criterion as follows.

Materials and Methods

Since new minimally invasive surgical techniques do not permit directly touching and palpating internal tissue, the surgeon remains unaware of its condition. For example, hard lumps in soft organs are detected by probing the tissue with fingers; arteries are localized during dissection by feeling for a time varying pressure; the structural integrity of a blood vessel wall is assessed by rolling it between the fingers.

Scrutinizing for the presence of an artery in soft biological tissue and distinction of a tumor was carried out to pattern out palpation by using the finite element method. It is also possible to separate a healthy artery from a stenotic artery by simulating palpation.

In every application of the tactile sensing method, physical contact between the tactile sensor and the object or tissue has special importance. In this physical contact, according to the sensor design, a parameter of touch was used as a criterion for stimulating a sensor. This criterion can be force, pressure (stress), displacement (strain), temperature, humidity, roughness, stiffness, and softness that appeared on the surface of the touched object.

In the present method, a phantom of soft tissue was considered that includes an artery and a tumor as illustrated in Fig. 10.16. It has been divided into four parts:

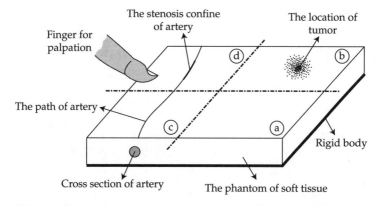

Figure 10.16 Schematic representation of a soft tissue: (a) the tissue itself (b) TIT (c) TIHA (d) TISA. [From Abouei (2008),[10] courtesy of Science Publications.]

1. The tissue itself
2. Tissue including a tumor (TIT)
3. Tissue including a healthy artery (TIHA)
4. Tissue including a stenotic artery (TISA)

The cross section of each part, as shown in Fig. 10.17, has been selected as a two-dimensional model. The cross section of model b and c pass through the center of the tumor and stenotic artery, respectively. For modeling of the palpation process using a finger, a constant displacement on the top line of each model has been considered; the von Mises stress parameter was selected as a detection criterion in modeling. For completing the simulation of palpation effect, the bottom line of each model was constrained in y-direction as shown in Fig. 10.17.

As shown in Fig. 10.17, four rectangles were considered as four plane models of the soft tissue, where each model measured 80 × 30 mm. In this modeling, two arteries and tumors were assumed to be completely circular and in the center of tissue geometry. The modulus of elasticity of the tumor was assumed to be twenty times larger than the tissue.

The aim of this modeling is to explore the effect of an artery, a stenotic artery, and a tumor in tissue and to compare them for finding a criterion for separating these issues. Therefore, the von Mises stress graph for all of nodes was created on the top side of each of the four models, shown in Fig. 10.17, and a two-dimensional tactile image of the models was also explored.

Results

For every model, outputs in two different situations were derived. There was a stress distribution on the top side of tissue model that

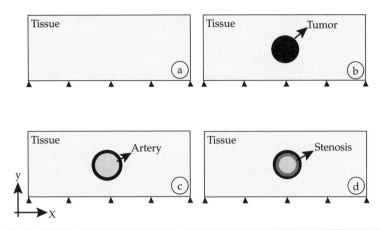

FIGURE 10.17 The cross section of each part shown in Fig. 10.16. [From Abouei (2008),[10] courtesy of Science Publications.]

is the site of touch between tissue and sensor. This was named a two-dimensional tactile image. Then, the von Mises stress graph on all of the top nodes was used. Based on the two-dimensional tactile imaging and stress graph, results were reached.

The results showed an appearance of the symptoms of existence of an artery or a tumor on the top side of tissue model. This result supports the suitability of the artificial tactile sensing method in either artery or tumor detection and separation of a healthy artery, stenotic artery, and tumor from each other. Figure 10.18 shows two-dimensional tactile images that correspond with the application of a constant displacement on the top line of each model. From the two-dimensional tactile images b, c, and d, it is understood that applied pressure on the tissue, which has a tumor or an artery embedded, caused a nonuniform stress distribution in comparison with the tissue itself.

The results also showed an appearance of a peak in stress graph. As shown in Fig. 10.19, Fig. 10.20, and Fig. 10.22, von Mises stress graphs derived for the nodes on the top side of models b, c, and d, in comparison to model a, include a peak. The peak for these three graphs indicates the existence of a tumor or an artery in the tissue. Also, since the centers of peaks are exactly under the tumor or artery, we can determine the exact location of the artery or tumor by viewing the stress graph.

In Table 10.3, for models TIHA and TISA, the von Mises stress has been presented for four systolic steps. From this, it can be deduced that for the TISA model, the variation range of von Mises stress in systolic steps from 0 to 0.425 seconds is smaller than the TIHA model, but the value of its von Mises stress is larger than the TIHA model.

Since pressure inside an artery increases linearly in systolic steps from 80 to 120 mmHg and decreases linearly in diastolic steps from

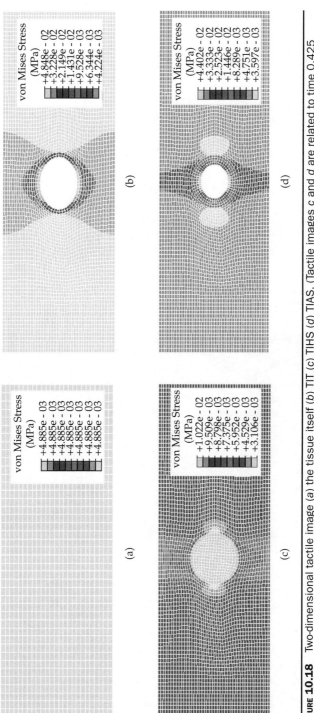

FIGURE 10.18 Two-dimensional tactile image (a) the tissue itself (b) TIT (c) TIHS (d) TIAS. (Tactile images c and d are related to time 0.425 second that internal pressure of artery reaches 120 mmHg.) [From Abouei (2008),[10] courtesy of Science Publications.]

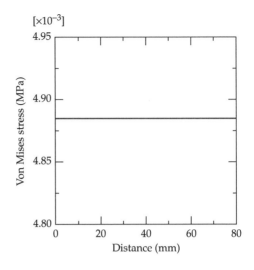

FIGURE 10.19 Von Mises stress graph for nodes on top side of model *a*. (This graph is the same during 0.85 second.) [From Abouei (2008),[10] courtesy of Science Publications.]

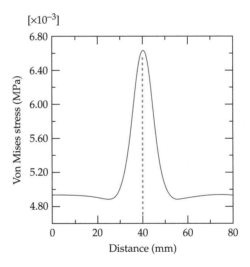

FIGURE 10.20 Von Mises stress graph for nodes on top side of model *b*. (This graph is the same during 0.85 second.) [From Abouei (2008),[10] courtesy of Science Publications.]

Model	Von Mises stress ($\times 10^{-3}$ MPa)			
	0.00 (s)	0.159 (s)	0.349 (s)	0.425 (s)
TIHA	5.550	5.740	6.102	6.254
TISA	6.281	6.322	6.400	6.432

Source: From Abouei (2008),[10] courtesy of Science Publications.

TABLE 10.3 The Values of von Mises Stress for Four Times in Systolic Step for Models TIHA and TISA.

120 to 80 mmHg, a graph can be derived similar to the graphs shown in Fig. 10.21 and Fig. 10.22 for diastolic step.

In this approach, the palpation of a physician was modeled and simulated using the finite element method. This modeling was performed for four two-dimensional tissue models: the tissue itself, tissue including a tumor, tissue including an artery, and tissue including a stenotic artery.

By comparing two-dimensional tactile images *a*, *b*, *c*, and *d* in Fig. 10.18, it can be seen that with applying same loading on the top side of each model shown in Fig. 10.17, then:

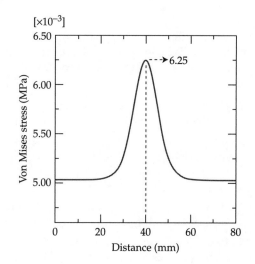

FIGURE 10.21 Von Mises stress graph for nodes on top side of model *c* at 0.425 seconds of systolic step. [From Abouei (2008),[10] courtesy of Science Publications.]

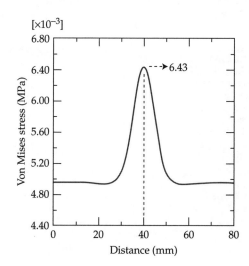

FIGURE 10.22 Von Mises stress graph for nodes on top side of model *d* at 0.425 second of systolic step. [From Abouei (2008),[10] courtesy of Science Publications.]

1. If the two-dimensional tactile image was uniform, as shown in Fig. 10.18a, there is only the tissue itself.

2. If the two-dimensional tactile image was not uniform, the tissue includes other material with different mechanical properties of soft tissue such as a tumor or artery, as shown in Fig. 10.18b, Fig. 10.18c, and Fig. 10.18d.

3. If the two-dimensional tactile image was not uniform but remained constant with time, the tissue contains a tumor, as shown in Fig. 10.18b.

4. If the two-dimensional tactile image was neither uniform nor constant, the tissue includes an artery, as shown in Fig. 10.18c and Fig. 10.18d.

By comparing von Mises stress graphs shown in Fig. 10.19, Fig. 10.20, Fig. 10.21, and Fig. 10.22, it can be concluded that, when applying the same loading on the top side of each model shown in Fig. 10.17, then:

1. If stress values versus top nodes and time is constant, as in Fig. 10.19, only the existence of the tissue is approved.

2. If the stress graph consists of a peak, the tissue includes other materials with different mechanical properties from soft tissue such as a tumor or artery, such as Fig. 10.20, Fig. 10.21, and Fig. 10.22.

3. If the value of stress peak versus time is constant, the tissue includes a tumor, such as in Fig. 10.21.

4. If the value of stress peak is time-dependent, the tissue includes an artery, shown in Fig. 10.23.

5. If we compare the peaks of two time-dependent stress graphs at the same time for a similar and proper loading, the peak that has

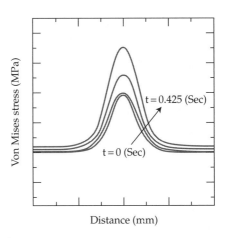

FIGURE 10.23
Von Mises stress graph for nodes on the top side of model of tissue including an artery at four different times of systolic step. [From Abouei (2008),[10] courtesy of Science Publications.]

larger value belongs to the stenotic artery because it has more strength than a healthy artery under the same conditions.

Consequently, based on the classifications presented in Table 10.4, the separation of a healthy artery from a stenotic artery is not possible by just comparing their two-dimensional tactile images; rather, they can be distinguished from each other by comparing their stress graphs.

10.6 Artificial Skin

To replicate the force-sensing capabilities of the human skin is still in its early stages. In fact, to develop such a pliable material has been the target of engineering research for many years. Application areas of this research include medical diagnoses, robotic manipulation, human-robot interaction, and haptic interfaces. For instance, we can refer to the application that involves achieving dexterous manipulation or exploration of the environment using a robotic hand. This task requires detailed knowledge of contact-forces between a manipulator and the interacting object. High-density tactile sensors on the robot would provide awareness of their environment when we are dealing with the area of human-robot interaction. Additionally, human injury can be prevented by using this system. Commercially available sensors suffer from various disadvantages. They do not have high receptor density, a multidimensional force discernment, and a flexible sensing array that can conform to arbitrarily shaped object, bodies or end effectors.[14,15]

There are three primary characteristics that are the main focus of research on artificial skin. They are high density, flexible sensing arrays, and the ability to decompose all three components of an applied force vector.[14] So far, some, but not all, of these characteristics have been accomplished by most research activities.

We now discuss a new flexible and high-density sensing array that provides tri-axial (i.e., three independent components) force discernment of an applied force for this array. Tekscan's piezoresistive

Model	Two-dimensional tactile image	Von Mises stress graph
Tissue itself	Uniform and time-independent	A constant value and time-independent
TIT	Nonuniform and time-independent	A peak and time-independent
TIHA and TISA	Nonuniform and time-dependent	A peak and time-dependent

Source: From Abouei (2008),[10] courtesy of Science Publications.

TABLE 10.4 The Comparison of Results for Four Models.

194   Chapter 10

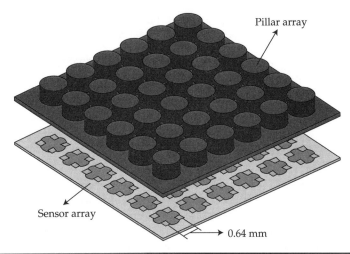

FIGURE 10.24 Sketch of the artificial skin design. The sensing elements are small square cells on the sensor array. Shaded outlines of the pillars are drawn on the sensor to visualize the alignment of the two arrays. (Koterba and Matsuoka, "A Triaxial Force Discernment Algorithm for Flexible High Density, Artificial Skin," *Proceedings of the 2006 IEEE International Conference on Robotics and Automation*, Orlando, Florida, May 2006, © IEEE.)

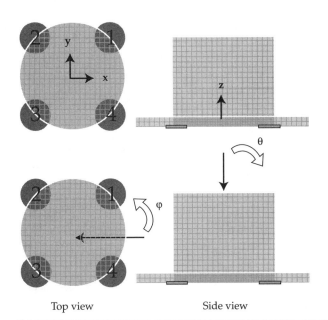

FIGURE 10.25 Top and side sketches of a single pillar. The darker, numbered regions are sensor locations and there are two different coordinate systems. (Koterba and Matsuoka, "A Triaxial Force Discernment Algorithm for Flexible High Density, Artificial Skin," *Proceedings of the 2006 IEEE International Conference on Robotics and Automation*, Orlando, Florida, May 2006, © IEEE.)

technology comprises the backbone of skin design. It is both thin and flexible. It also allows high-resolution arrays, and provides excellent sensitivity. This prototype incorporates Tekscan's #5027 sensors. The specifications are a pressure resolution higher than 0.02 kg/cm^2 and a spatial resolution of 250 "sensels" per square centimeter.[16] It should be noted that three separate components of the applied force cannot be provided by Tekscan sensors alone. This problem is tackled by an array of small polydimethylsiloxane (PDMS) pillars. As shown in Fig. 10.24, these pillars are added on top of the sensing array. Any force applied to the top of the pillar is directed down through the base of the pillar. At the same time, the applied force is concentrated around the outer edge of it. Four sensors are placed at the base of each pillar, as seen in Fig. 10.25, centered under the edge. By so doing, we can obtain maximum response. To decrease the applied force vector on the corresponding pillar, the data from each element are combined.

Besides delivering triaxial force information, the PDMS pillars can provide us with compliance and surface friction. In other words, they can act similar to human skin.[14]

References

1. L. Gao, K.J. Parker, R.M. Lerner, and S.F. Levinson, "Imaging of the Elastic Properties of Tissue—A Review," *Ultrasound in Medical & Biology*, vol. 22, no. 8, 1996, pp. 959–977.
2. A. Vinckier and G. Semenza, "Measuring Elasticity of Biological Materials by Atomic Force Microscopy," *FEBS Letters*, vol. 430, no.1–2, 1998, pp. 12–16.
3. Y. Uchio, M. Ochi, N. Adachi, K. Kawasaki, and J. Iwasa, "Arthroscopic Assessment of Human Cartilage Stiffness of the Femoral Condyles and the Patella with a New Tactile Sensor," *Medical Engineering & Physics*, vol. 24, 2002, pp. 431–435.
4. K. Miyaji, S. Sugiura, H. Inaba, S. Takamoto, and S. Omata, "Myocardial Tactile Stiffness During Acute Reduction of Coronary Blood Flow," *The Annals of Thoracic Surgery*, vol. 69, no. 1, 2000, pp. 151–155.
5. G. Darb Emamieh, A. Ameri, S. Najarian, and A. Tavakoli Golpaygani, "Experimental and Theoretical Analysis of a Novel Flexible Membrane Tactile Sensor," *American Journal of Applied Sciences*, vol. 5, no. 2, 2008, pp. 122–128.
6. C. Shan-lin and Z. Zhou-lian, "Large Deformation of Circular Membrane Under the Concentrated Force," *Applied Mathematics and Mechanics*, vol. 24, no. 1, 2003, pp. 28–31.
7. A. Tavakoli Golpaygani, S. Najarian, M. Mehdi Movahedi, and G. Darb Emamieh, "Fabrication of a Capacitance-Based Tactile Sensor with Biomedical Applications," *American Journal of Applied Sciences*, vol. 5, no. 2, 2008, pp. 129–135.
8. Y. Hasegawa, M. Shikida, T. Shimizu, T. Miyaji, H. Sasaki, K. Sato, and K. Itoigawa, "A Micromachined Active Tactile Sensor for Hardness Detection," *Sensors and Actuators*, vol. 114, 2004, pp. 141–146.
9. H. Roham, S. Najarian, S. M. Hosseini, and J. Dargahi, "Design and Fabrication of a New Tactile Probe for Measuring the Modulus of Elasticity of Soft Tissues," *Sensor Review*, vol. 27, no. 4, 2007, pp. 317–323.
10. A. Abouei Mehrizi, S. Najarian, M. Moini, and F. Tabatabai Ghomshe, "Tactile Distinction of an Artery and a Tumor in a Soft Tissue by Finite Element Method," *American Journal of Applied Sciences*, vol. 5, no. 2, 2008, pp. 83–88.

11. P. Dario and M. Bergamasco, "An Advanced Robot System for Automated Diagnostic Tasks through Palpation," *IEEE Transactions on Biomedical Engineering*, vol. 35, no. 2, 1988, pp. 118–126.
12. W. J. Peine, J. S. Son, and R. D. Howe, "A Palpation System for Artery Localization in Laparoscopic Surgery," First International Symposium on Medical Robotics and Computer-Assisted Surgery, Pittsburgh, 1994, 22–24.
13. R. A. Beasley and R. D. Howe, "Tactile Tracking of Arteries in Robotic Surgery," *Proceedings of the 2002 IEEE International Conference on Robotics & Automation*, Washington, DC, 2002.
14. S. Koterba and Y. Matsuoka, "A Triaxial Force Discernment Algorithm for Flexible High Density, Artificial Skin," *Proceedings of the 2006 IEEE International Conference on Robotics and Automation*, Orlando, Florida, May 2006.
15. D. Yamada, T. Maeno, and Y. Yamada, "Artificial Finger Skin Having Ridges and Distributed Tactile Sensors Used for Grasp Force Control," *Journal of Robotics and Mechatronics*, vol. 14, no. 2, 2002, pp. 140–146.
16. http://www.tekscan.com/industrial/catalog/5027.html

CHAPTER 11
Application of Tactile Sensing in Robotic Surgery

11.1 Robot Definitions

The three definitions of a robot are given as follows:

The Robot Institute of America (1979) offers the definition of a robot as "a reprogrammable multifunctional manipulator, designed to move material, parts, tools or specialized devices through variable programmed motions for performing a variety of tasks." The second definition, which is more applicable to surgical procedures is "a powered computer controlled manipulator with artificial sensing that can be reprogrammed to move and position tools to carry out a range of surgical tasks."[1] The third definition is "robotic systems for surgery are firstly computer-integrated surgery (CIS) systems, and secondly medical robots." In other words, the robot itself is just one element of a larger system designed to assist a surgeon in carrying out a surgical procedure; this procedure may include preoperative planning, intra-operative registration to presurgical plans, use of a combination of robotic assistance and manually controlled tools for carrying out the plan, and postoperative verification and follow-up.[2]

An Aspect of an Integrated System

As mentioned earlier, one can use a robot in order to conduct a surgical procedure. For example, in a knee surgery performed by a robot, typical stages described in Table 11.1 must be performed. To achieve this, certain stages can be followed. These can be categorized as preoperative, intra-operative, and postoperative.[1] Another stage can also be included depending on whether or not there is a need for further cuts.

Stages	Description
Pre-operative	Image patient
	Edit images and create three-dimensional model of leg
	Create three-dimensional model of prostheses
	Superimpose prostheses over three-dimensional model of leg
	Adjust and optimize location
	Plan operative procedure
Intra-operative	Fix and locate patient on table
	Fix and locate robot (on floor or on table)
	Input three-dimensional model of cuts into robot controller
	Datum robot to patient
	Carry out robot motion sequence
	Monitor for unwanted patient motion
Postoperative	Remove robot from vicinity
	Release patient
	Check quality of procedure
If further cuts are necessary	Reclamp patient
	Reposition and datum robot to patient
	Repeat robotic procedure

Source: From Davies (2000),[1] courtesy of Professional Engineering Publishing.

TABLE 11.1 Typical stages in robotic knee surgery.

11.2 Application of Robots in Surgery

The main advantage of robotic technology is to improve the quality of a typical surgical procedure. This is obtained by a noticeable improvement in various features such as dexterity, precision, and stability. If one uses image-guided techniques, robotic surgery will benefit from MRI and CT-scan data by directing various instruments towards the site of operation.[3] By doing this, we will need new algorithms, user interfaces for planning procedures, and also specialized sensors. These types of sensors serve as tools for registering the patient's anatomy with the preoperative image data obtained in the earlier stages. In MIS, the surgeons can perform various procedures without making large incisions inside the patient's body. Here, the surgeon can remotely control the surgical robot. However, due to the loss of tactile feedback and the access constraints one encounters, new designs and sensing technologies are required. This will enhance the surgeon's dexterity.

The surgical robots have a considerable number of applications, including neurosurgery. Here, an image-guided robot can take samples from brain lesions. Using careful controlling, there will be less damage to tissues surrounding the lesions. Another example is orthopedic surgery. In this procedure, the role of the robot is to shape the femur so that it can precisely fit prosthetic hip joint replacements. Other examples are closed-chest heart bypass surgery, various procedures in ophthalmology, and surgical training and simulation.[3] So far, robotic surgery has received very promising feedback. However, various concerns, including safety, clinician acceptance, performance validation, and high capital costs, need to be investigated in more details.

It is obvious that there are many differences between machines and humans when it comes to their features and capabilities. In many aspects, we can see noticeable benefits when using surgical robots. A summary of some of the advantages and disadvantages of surgery using machines and humans is given below.[4]

The advantages of human performances include: strong hand-eye coordination; good dexterity (at a human scale), flexibility and adaptability; an ability to integrate extensive and diverse information; an ability to use qualitative information; good judgment; and an ability to instruct and debrief by making a report for someone. The limitations of human performances are: limited dexterity outside the natural scale; a propensity to tremor and fatigue; limited geometric accuracy; limited ability to use quantitative information; the need for large operating room space; limited sterility; and susceptible to radiation and infection. The advantages of robotic performances are: good geometric accuracy; stable and untiring; designed for a wide range of scales; can be sterilized; resistant to radiation and infection; and an ability to use diverse sensors (chemical, force, and acoustic) in control. The limitations of robotic performances are: poor judgment, limited dexterity and hand-eye coordination, limited to relatively simple procedures, expensive, technology influx, and, finally, difficult to construct and debug.

Perhaps, the main difference between these two is precision and accuracy.[3] Machines can normally employ a considerable amount of detailed data. In this regard, robots can guide the surgical instruments accurately to the treatment site. They accomplish this by combining 3D imaging data, computers, and various sensors. Surgical robots have another feature that clearly favors them when they work through incisions. Here, we are dealing with incisions that are much smaller than the size of human hands. Therefore, at these small scales, the tremor of the human hand causes serious limitations and, thereby, affects the outcome of the surgery.[3]

Robotics in Surgery

In recent years, the use of robots in surgery has rapidly expanded. These machines can improve precision, filter human motion tremor, extend human reach into the body, and reduce the risk of infection.[5]

One type of classification for surgical robots is based on the planning strategy, and includes model-based robotic surgery and non-model-based robotic surgery.[5]

Surgical robots can also be classified based on the level of surgical invasiveness which was discussed in Chap. 7, Sec. 7.1. This section looked at open surgery and minimally invasive surgery.

In model-based robotic surgery, we use various geometric models. These models can be obtained from scanned images using computer tomography (CT) or magnetic resonance imaging (MRI) during the preoperation stage.[5] Here, we use a process which is called *registration*. During a registration process, the preoperative model is matched with an intra-operative model. The robotic surgery becomes an off-line robot motion planning problem (as in a CAD-CAM system) if the registration process can be accurately performed.[5] Applications for this method are surgeries of hard tissues such as spinal surgery, neurosurgery, hip replacement surgery, total knee replacement surgery, plastic surgery, and eye surgery. The accuracy of the model has a major impact on the quality of the surgery.

In non-model-based robotic surgery, we typically deal with soft tissues. In these cases, the model is either not available or not useful during operations. This is due to the tendency of tissues floating in the body or changing in shape. Non-model-based systems are primarily being used during surgeries on a body organ.[5] Surgeons either navigate a hand-held robot directly, or manipulate an input device in the case of teleoperation. Here, the man/machine interface plays a key role since surgeons are directly responsible for manipulating the surgical robots. In non-model-based robotic surgery, the surgeon's expertise is the key factor which affects the quality of surgery.[5] Additionally, the surgical results can improve by guiding surgeons using force feedback and cooperative control.

Current Applications of Robotic Surgery

Current applications of robotic surgery are categorized in Fig. 11.1.

1. Orthopedic surgery: One of the major applications of robotic surgery is in orthopedics surgery, especially in total hip arthroplasty. In digital surgeries of total hip and total knee arthroplasty (THA and TKA), for preoperative planning, we use computer navigation systems. This technique is accurate and efficient when dealing with bones.[6] Unlike soft-tissues, bones are less prone to deformation when pressed upon. Therefore, we can utilize software based on the rigid-body assumption with sufficient accuracy.

2. Neurosurgery: Improvements in computer technology, engineering, minimally invasive surgery, along with the new neuroimaging techniques, have created the concept of digital robotic neurosurgery.[6] High precision in brain surgeries are

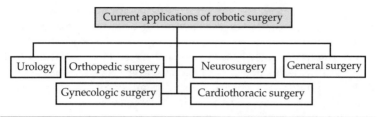

FIGURE 11.1 Categories of current applications of robotic surgery.

necessary in order to localize and manipulate within the brain together with the relatively fixed landmarks of the cranial anatomy. For this reason, neurosurgery was one of the first organ systems where robotic surgery was introduced.

3. Cardiothoracic surgery: Recently, many coronary artery problems have increasingly been treated by angioplasty. This method is, in fact, a type of minimally invasive procedure. Other options that we might be compelled to use are coronary artery bypass grafting (CABG) or off-pump coronary artery bypass grafting (OPCAB).[6] The main disadvantage of using these options was the need for employing open chest surgery. However, with the introduction of robotics, surgeons are now able to replace sternotomy and thoracotomy with small incision entries.

4. Urology: Since the late 1980s, a number of benefits have been demonstrated by industrial robots that have encouraged urological surgery to embrace the usage of robotics for health care delivery. The benefits included long-term economic advantage, increased accuracy, and improved quality. The first urological robot, also known as URobot, was a PROBOT in 1989. It was used in clinical trials for transurethral resection of the prostate (TURP).[6]

Suturing in MIS

Suturing has a long history. It started from closing the wound with an ant's head in ancient times to today's laparoscopic stitching with absorbable materials. It is still one of the most difficult tasks, and accounts for a significant percentage of operating time in MIS. Note that the ant itself has very strong mandibles that can serve as a suture.[5] It is left to bite close to the wound. Following this, its head is twisted off and its pincers (tool for gripping things) holds the wound tightly closed. Finally, the ant's saliva causes the skin to swell, thus closing the edges of the wound.

Currently, there is ongoing research on alternatives to suturing techniques such as tissue gluing, stapling, and thermal tissue bonding.

However, suturing is still the primary tissue approximation method. For example, in the laparoscopic suturing task, we demand a high level of surgeon's skills and endurance.[5] Despite the needs to perform suturing autonomously, no major research efforts have been published. It is worthwhile to decompose the suturing task into the subtasks and figure out what technical difficulties are encountered with each subtask.[5] There are several knot-tying techniques in surgery and two of them (square knot and surgeon's knot) are illustrated in Fig. 11.2.

1. Square knot (reef knot): This is the easiest and most reliable method for all suture materials which consists of two half knots in opposing directions.
2. Surgeon's knot: This knot requires an additional loop in the first throw increasing the surface contact of the suture.[5] It results in doubling the friction for the initial half knot, and is, therefore, more stable than the square knot. This knot is generally used in open surgery, and is sometimes referred to as a *friction knot*.[5]

The square knot is the most common knot technique in surgery. Since it requires fewer steps than the surgeon's knot, it is an ideal knot for laparoscopic surgery. Based on the observations of the manual suturing operations, the suturing task can be broken down into the following subtasks:[5] stitch, create a suture loop, develop a knot, place a knot, and secure a knot.

It should be noted that, once the knot is completed, a sufficient amount of tension must be applied to tighten the knot. However, this tension should not be so much as to damage the tissue. With the laparoscopic instruments such as long sticks, it is very hard to apply the precise tension. We have to be careful not to tear off the soft tissue or loosen the knot. The technical difficulty involves regulating the tension along the desired direction.

Laparoscopic Suturing

Suturing is one of the most difficult tasks in various surgeries. It takes a significant percentage of operating time and includes a complex motion planning. There is an extended learning curve that surgeons

Half knot   Square knot   Surgeon's knot

FIGURE 11.2 Typical knot tying techniques. [From Kang (2002),[5] used by permission.]

must go through to gain the required skill and dexterity.[5] This is a direct result of constraints in laparoscopic surgery. To add to this complexity, we observe a great deal of variability, even among trained surgeons. Time and motion studies of laparoscopic surgery have indicated that, for operations such as suturing, knot tying, suture cutting, and tissue dissection, the operation time variation between surgeons can be as great as 50%.[5] It was noted that the major difference between surgeons in suturing lies in the proficiency at grasping the needle and moving it to the desired position and orientation. Furthermore, they should be careful not to slip or drop the needle. The motion analysis on the suturing task using conventional needle holders was performed by Cao et al.[7] The task was broken down into five basic motions such as reach and orientation, grasp and hold, push, pull, and release. The teleoperator slave system, with a dexterous wrist for minimally invasive surgery, was developed by Madhani et al.[8] They demonstrated the suturing task with direct vision. Obviously, autonomous robotic suturing is a challenge in robotic surgery when direct vision is limited.

The primary difficulties in autonomous robotic suturing are: manipulating the curved needle; dealing with a flexible suture; getting a trajectory for knot placement; and applying the proper tension.[5]

Tension Measurement in Suturing

It is very difficult to measure a tension in the suture in minimally invasive surgery. This is due to the geometric constraints on mounting a sensor and to sterilizing issues.[5] For force measurements, strain-gauge sensors are most widely used. By mounting them on the laparoscopic instruments, we can measure the strains of the instruments due to the external forces. This is shown in Fig. 11.3. The main features of these sensors are high sensitivity, accuracy, and bandwidth with a simple electronic circuit. The tension can be measured directly using these sensors. The main drawback of using them in MIS is that they are very sensitive to temperature variation.[5] As a result, any contact

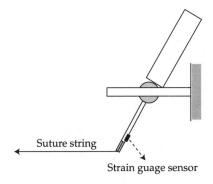

FIGURE 11.3
Strain gauge transducer. [From Kang (2002),[5] used by permission.]

Suture string

Strain guage sensor

with soft human tissue during the surgery may cause a drift in the outputs. Another problem is that, in order to be sterilized for repeated use, surgical instruments are subjected to a high temperature process. Therefore, we will need to use disposable strain gauge sensors. We should also consider the calibration of each tool.

Vision sensors can be used with premodeled and premarked sutures. As shown in Fig. 11.4, if the sutures have marks at a predetermined spacing, the tension can be estimated by measuring the elongation between two marks. The difficulty with this method is that the marks may not be visible.[5] This may happen due to the contamination of blood or the occlusion of the sutures for the instruments. It is often difficult to have a high bandwidth and resolution.

As depicted in Fig. 11.5, the way to measure the tension is to put a force/torque sensor in the base of the manipulator, such as a base sensor. This has two main advantages; the sensor does not need to be sterilized, and force measurement (and hence control) is independent of the tools.[5]

In addition to these benefits, the sensor information can be used to compensate for the friction on the joints for precise motion control. The major problem with this method is that the sensor outputs have configuration dependent gravity and motion induced components.[5] Precise dynamic models and the dynamic parameters are required so that we can compensate for the dynamic and static effects of the signals.

FIGURE 11.4
Vision sensor. [From Kang (2002),[5] used by permission.]

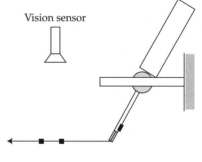

FIGURE 11.5
Base force/torque sensor. [From Kang (2002),[5] used by permission.]

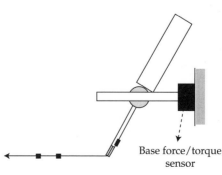

Commercial Robots for Surgery

In this section, we will first give a summary of the activities of some companies which produce commercial robots. We will discuss the surgical application of their products, and will then go into more detail about the practical uses of commercial robots for surgery.

Companies Which Produce Commercial Robots

Computer Motion Computer Motion is a high-tech medical device company. This company played a very important role in developing early surgical robotic technology. They specialized in medical robotics, and they developed a baseline for integrated robotic surgery. The company designed and manufactured systems for the Intelligent Operating Room of the future. Additionally, they made a significant contribution to the field of minimally invasive surgical robotics.[9]

Their first product was AESOP, a robotic system used to hold an endoscope camera. It has great application in minimal invasive laparoscopic surgery. Later on, an extension of AESOP's arms was marketed in order to control surgical instruments. This was the ZEUS Robotic Surgical System with three robotic arms attached on the side of the operation table. Another important robotic system is HERMES. Although it does not use robotic arms, it is used to connect all the intelligent tools in the operating room.

Intuitive Surgical In 1995, another strong competitor in the field of robotic surgery, Intuitive Surgical, was founded based on foundational robotic surgery technology developed at the Stanford Research Institute (SRI) International.[10] da Vinci is a famous product of this company.

Integrated Surgical Systems Incorporated Another medical robotics company, Integrated Surgical Systems Inc., specialized in surgical robots for hip and knee implants. The Orthodoc and Robodoc designed for orthopedic surgery was first developed in a collaborative project between the University of California, Davis, and the IBM T. J. Watson Center.[9] NeuroMate is used for surgical assistance for biopsy and tumor removal. It is a neurosurgery system developed by MMI in combination with Robodoc.

Accuray Incorporated Radio-surgery is also incorporating advanced robotic systems. To achieve this, CyberKnife was developed by Accuray Incorporated in 2001. Using radiation, it provides lesion treatments anywhere in the body.

Commercial Robots for Surgery

As mentioned, typical commercial robots for surgery were categorized as da Vinci, ZEUS, AESOP, Robodoc, Orthodoc, and NeuroMate. Here, we discuss the practical application of each of these models.

da Vinci There are four main components to da Vinci: the surgeon console; the patient-side cart; the EndoWrist instruments; and the Insite Vision System with high-resolution three-dimensional endoscope and related image-processing equipment.[9] When using this system, the surgeon is situated at this console several feet away from the patient on the operating table, as seen in Fig. 11.6. The surgeon has his head tilted forward and his hands inside the system's master interface. The surgeon sits viewing a magnified 3D image of the surgical field with a real-time progression of the instruments as he/she operates.[10] The instrument controls enable the surgeon to move within a one cubic foot area of workspace.

AESOP (A voice-controlled endoscope positioning robot) AESOP (Automated Endoscopic System for Optimal Positioning) approximates the form and function of a human arm. It also allows control of the endoscope through simple verbal commands. Thus, it eliminates the need for a member of a surgical staff to manually control the camera, while providing a more stable and sustainable endoscopic

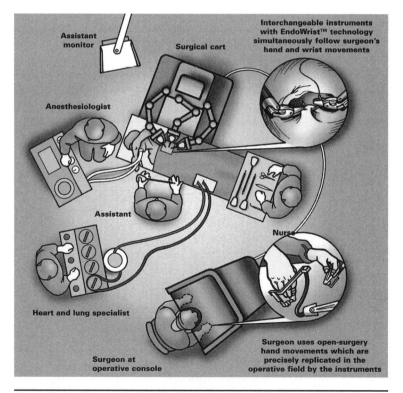

Figure 11.6 Schematic of performing suturing task by master-slave robot. (Photo courtesy of Intuitive Surgical, Inc., 2008.) (See also color insert.)

image. It is reported that the AESOP's robotic arm holds the endoscope with a steadiness that no human can match.

ZEUS Surgical System   The ZEUS Surgical System is made up of an ergonomic surgeon control console and three table-mounted robotic arms. It performs surgical tasks and provides visualization during endoscopic surgery. The surgeon, while seated at an ergonomic console with an unobstructed view of the operation room, can control the right and left arms of ZEUS. This translates to real-time articulation of the surgical instruments. A third arm incorporates the AESOP Endoscope, providing the surgeon with magnified, rock-steady visualization of the internal operative field.[11]

HERMES Control Center (a humanoid service robot)   Hermes Control Center is actually a centralized networking platform. It is designed to control many devices which allow surgeons to voice-control them. By networking all the components in the room, it is also possible to provide the robotic system with more control in the operating room environment. These include such things as changing positions of a table, video cameras, surgical equipment, and lighting conditions.[9]

Robodoc   Robodoc is a computer-controlled surgical robot. It is equipped with specialized drilling components and other hardware for preparing bones for prosthetic implants. The robot drills cavities for hip implants, removes bone cement for revision surgeries, and planes surfaces on the femur and tibia for knee implants. This surgical robot is capable of performing primary total hip replacement, revision hip replacement, and total knee replacement surgeries.[12]

Orthodoc   To plan precisely for the surgery, the Orthodoc preoperative planning workstation is used. Using Orthodoc, the surgeon can simulate the surgery using the actual CT-scan of the patient. The advantage of using this system is that it allows the surgeon to select an optimal implant for the patient. By doing this, we can reduce material waste and procedural costs. This system also allows the surgeon to review the details of the patient's anatomy and determine how best to position the implant.[12] By using the total knee planning module, which runs on the Orthodoc presurgical planning system, the patient's CT-scan data can be employed to provide the surgeon with a more precise plan of the surgery. Additionally, total hip planning is possible using this system.

NeuroMate   For stereotactic functional brain surgeries, the surgeons can use NeuroMate. This is an image-guided, computer-controlled, robotic system. Additionally, it includes a presurgical planning workstation. The NeuroMate System positions, orients, and manipulates the operating tools within the surgical field. This is done exactly as planned by the surgeon on the image planning workstation.[12]

One of the greatest challenges facing surgeons who were training on these devices was that they felt hindered by the loss of tactile. In other words, they did not have haptic sensation or ability to feel the tissue.

11.3 Robots for MIS

Robotic and mechatronic systems are key technologies to overcome the drawbacks of manual MIS. They help the surgeon to virtually regain direct access to the operating field. Mechatronic systems are characterized by a high level of integration mechanics, electronics, and software into one optimized technical unit. A combination of mechatronic systems with telepresence and telemanipulation approaches lead to minimally invasive robotic surgery (MIRS) which is a subgroup of MIS.[13] We can achieve *telepresence* if the human operator is provided with the impression of actually being present in a remote environment. *Telemanipulation* means that the operator is not only present passively, but also able to interact actively with the remote environment. MIRS telepresence/telemanipulation systems help the surgeon to overcome barriers, such as the patient's chest or abdominal wall. The undesired reverse hand motion can be avoided using appropriate control algorithms. Another advantage is the downscaling of the surgeon's hand motion before it is transmitted to the robot. Here, movements of instruments become more accurate than in endoscopic surgery. Another benefit of this system is that, by using low-pass filters, the surgeon's tremor can be decreased.[13] Currently, the commercially available MIRS systems are Vinci and ZEUS.

Force Sensors for Surgical Robots

Force feedback is an obvious feature missing from current robotic surgical systems. Many surgeons would like to have this feature in their instruments. However, it remains unimplemented in surgical systems. The problem is with the ability to accurately sense the interaction forces between the instrument and the internal organs and tissues of the human body.

The leading surgical robot, the da Vinci, has the capability of recreating forces against the hand. However, no attempt has been made, so far, to implement force feedback at the instrument tips, as this has proven to be a difficult task. This difficulty is somewhat related to the surgical environment. It is known that this environment is quite harsh and imposes special design requirements. The force sensor needs to fit through a small port (5 mm to 12 mm) and transmit the sensed force information back outside the body. For general procedures, the force sensor should sense up to 5 N of force.[14] In most designs, the sensor is located on the distal end of the instrument. This is due to the friction forces between the instrument and the port.[14]

The use of cable drives for articulated instrument wrists causes an additional difficulty. To improve the current designs, the surgical instrument should be sterilizable, waterproof, and impervious to temperature differences. The last feature needs to be incorporated since we change the environment of the tool from room temperature to internal body temperature.

Some of typical applications of force feedback in robotic surgery include:

- Manipulation of soft tissue (e.g., tissue grasping)
- Suturing (e.g., knot tying task with fine suture)
- Tissue cutting/spreading (e.g., dissection)
- Tissue recognition
- Needle insertion
- Detection of embedded object in tissue (e.g., lump detection)

As mentioned above, dissection is one important example of force feedback in robotic surgery. Dissection means cutting, separating, and examining parts of an animal or plant specimens for scientific or medical study. It is composed of three distinct phases: tissue recognition; accurate instrument positioning; and tissue cutting/spreading. During the dissection, the surgeon tries to minimize tissue trauma and preserve the surrounding structures.[14]

Teleoperation

Typical telerobotic systems consist of two major components, a slave manipulator and a master controller. These two units are separated by some distance. They could be located in different rooms of a hospital or even in different countries. The separation between these two units allows the operator to interact with an environment that would, otherwise, be too impractical, or even dangerous, such as nuclear plants. Due to the safety issues, human presence is not feasible in certain locations. Some examples include under-water exploration, hazardous material handling, and space exploration.[15]

The above systems have been in use for many years. Another usage of telerobotic systems, aside from hazardous environments, is to provide an amplification of the operator's input. With this, the operator is allowed to perform tasks that are beyond human strength limitations. For instance, a digging machine makes use of this technique. Maneuvering in small spaces where human presence is not possible is another application of teleoperation systems.[15] The inspection of pipes is an example of telepresence in an inaccessible location. Laparoscopic surgery is an example in which biomedical engineers have provided surgeons with the necessary teleoperation tools. In these types of surgeries, using smaller incisions together with a vision system and small robotic tools, a surgeon is able to operate on a treatment site.[15]

In fact, telesurgery is an integration of multimedia, telecommunications, and robotic technologies.[15] Here, the surgical procedures can be performed at a distance utilizing only a master and a slave manipulator. As shown in Fig. 11.7, the surgeon, who is sitting at the console, has control over the robotic arms, but the surgeon does not operate on the patient directly. Minimally invasive surgeries are the main application areas for telesurgical procedures.

Like any other modern systems, telesurgery faces a number of challenges. For instance, the time delay between master and slave sites is an especially challenging issue for all telesurgical applications.[16] The computation delay ranges up to 10 ms induced by computer networks. If we use multiple satellite communication links, however, the delays will be only a few seconds. Various disadvantages of these delays could be: dangerous instability of operation, loss of teleoperator fidelity, and loss of task performance.[17]

An interesting application of teleoperation is in treating wounded people in battlefields. As discussed above, since the surgeon can be anywhere in the world when controlling the robotic surgical system, it can be very useful for this application. The delivery of healthcare services by all healthcare professionals is very important, where distance is a critical factor. To tackle this, we can use the concept of *telemedicine*. This approach is basically the exchange of real-time data of medical information between physicians in different locations. In fact, it is considered as a rapidly developing application of clinical medicine where medical information is transferred via various methods. These methods could be telephone, the Internet, or other networks for the purpose of consulting. Telemedicine is even, at times, used for remote medical procedures or examinations. In general, telemedicine refers to the use of communications and information technologies for the delivery of clinical care.[18] It can be expressed in three parts:

1. The patient and doctor are located in different places.
2. The patients can be examined, treated, and monitored.

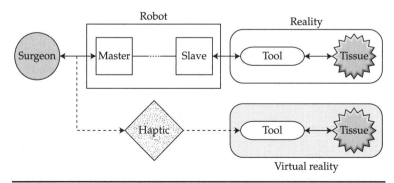

FIGURE 11.7 Flowchart of MIS by master-slave robot.

Application of Tactile Sensing in Robotic Surgery

3. The patient's data (text, voice, images, or even video) can be sent from a remote location, and medical advice can be offered from a specialty center.

Telemonitoring Skin Conditions

Skin Lesions A skin lesion is a superficial growth or patch of the skin that does not resemble the area surrounding it. Some different types of skin lesions are presented in Fig. 11.8. They are papule, nodule, wheal, macule (or macula), vesicle, pustule, and anetoderma.[19]

As represented in Fig. 11.9, in anetoderma, the soft subcutaneous tissue can be easily compressed through the dermal defect. Neurofibromas may also have a similar morphology, with a

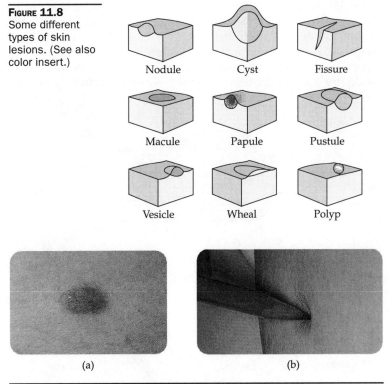

FIGURE 11.8 Some different types of skin lesions. (See also color insert.)

FIGURE 11.9 Appearance of anetoderma in (a) normal conditions and (b) when being touched. (Reprinted from White and Cox, *Diseases of the Skin:* Chapter 22, "Structural Disorders of the Skin and Disease of Subcutaneous Tissues; Structural Disorders of the Dermis, Collagen, and Elastic Tissue," Copyright © 2006, with permission from Elsevier.) (See also color insert.)

dumbbell-shaped soft nodule traversing the dermis, and capable of some degree of compressibility.[20]

Haptic Sensor for Monitoring Skin Conditions The sensor shown in Fig. 11.10 is concerned with the development of a haptic sensor that evaluates the human skin condition. The base of the sensor is an aluminum shell, around which are: a rubber sponge layer; PVDF film; a protective layer of acetate film; and gauze (finely woven fabric) that are stacked in sequence.[21] The sensor is attached on the tip of an elastic cantilevered beam and pressed against the surface of the skin. It is then slid over the skin to collect the surface morphological features. The sensor is applied to view the variation of skin conditions depending on age. A schematic of the experimental setup for this sensor is shown in Fig. 11.11.

A Tactile Sensor for Detection of Skin Surface Morphology

In the 1990s, serious interest in this method of healthcare delivery emerged. This was mainly due to improvements in the available technology, accompanied by falling transmission costs.[22] Many specialists, ranging from image-dependent areas such as radiology and pathology to traditionally non-image-dependent areas such as psychiatry and surgery, have found significant utility in telemedicine

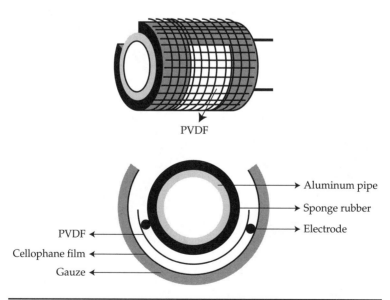

FIGURE 11.10 Schematic diagram of PVDF piezofilm sensor. (Reprinted from *International Journal of Applied Electromagnetics and Mechanics*, vol. 14, Tanakaa et al., "Haptic Sensor for Monitoring Skin Conditions," pp. 397–404, 2001, with permission from IOS Press.)

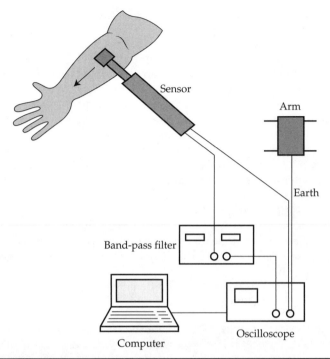

FIGURE 11.11 Experimental setup. (Reprinted from *International Journal of Applied Electromagnetics and Mechanics*, vol. 14, Tanakaa et al., "Haptic Sensor for Monitoring Skin Conditions," pp. 397–404, 2001, with permission from IOS Press.)

applications.[23] Although teledermatology is one of the best studied disciplines in telemedicine, several of these issues remain unresolved and require further investigation.[24]

At present, teledermatology is practiced in three different methods: store-and-forward (SAF); real-time (RT); or videoconferencing with a combination of the two techniques.[22] The SAF variant uses asynchronous data transfer technology, such as email, while RT teledermatology is based on synchronous data transfer technologies, such as videoconferencing software.[25] Real-time video conferencing has been shown to be an effective substitute for the face-to-face consultation; but it is both time-consuming and expensive. With the advent of digital cameras, store-and-forward systems with a history and digital image sent by email for an opinion have been found to be accurate.[26]

In addition to legal issues and the problems related to acceptability of teledermatology systems, these systems are likely to have some technical problems. One of the major problems is the lack of tactile sensing due to the available distance between the patient and the dermatologist.[27] Dermatology is primarily a visual subject; however, other sensory modalities, especially touch, and sometimes smell for anaerobic infection, are also important.[28] Skin surface morphology, which reflects its surface characteristics, such as softness, roughness, and the surface profile, has a great importance in diagnosis of dermal diseases. Most of these properties are not recognized just by the sense of vision. The most important problem mentioned by the physicians in the implementation of a calibrated teledermatology system was the inability to feel the skin.[29] Despite the importance of sense of touch in dermatology, present teledermatology systems are based solely on the photographs and films that are taken from the patient.[30,31]

A new type of tactile sensor was proposed that is capable of determining the surface morphology of skin lesions with application in telemedicine systems, including teledermatology.[27] The tactile sensor system consists of a tactile sensor, a data acquisition device, and a personal computer for data analysis. The sensor is moved on the surface of the models of skin lesions that are made from paraffin gel (as soft tissue or skin). Data from the sensor is collected and sent to a PC where the surface profile of the lesion will be extracted. By comparing the surface profile of different lesions, a dermatologist can make a more reliable diagnosis.

Tactile Sensor Structure The tactile sensor operation is based upon Faraday's induction law. According to this law, if a coil is placed in the varying magnetic field produced by another coil, a voltage will be induced in it which is proportional to the number of its turns and the rate of change of magnetic flux in it. This is given by Eq. (11.1):

$$e = N\frac{\partial \Phi}{\partial t} \qquad (11.1)$$

Where, N is the number of turns of the coil, $\partial \Phi / \partial t$ is the rate of change of magnetic flux, and e is the induced voltage. The amount of flux passing the coil is related to the core material and the position of the core with respect to the coil.

The tactile sensor is made up of one primary coil and two secondary coils, shown by letters A and B in Fig. 11.12. The primary coil is excited by a sinusoidal voltage that generates a varying magnetic field; this causes a voltage to be induced in each secondary coil. These voltages, which are shown by V_a and V_b in Fig. 11.12, are in phase but their magnitude is not equal and is calculated from Eq. 11.1. Since the numbers of turns of the secondary coils are equal, the induced voltage is solely proportional to the position of the core. At the center (where an equal length of core is inside each coil), the induced voltages are equal. When the core moves up, the voltage of the secondary coil A (V_a) will become greater than the voltage of the secondary coil B (V_b), and the difference of these voltages ($V_a - V_b$) is positive in sign. When the core moves down, the sign of the voltage

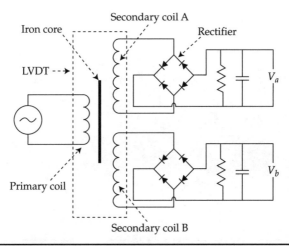

FIGURE 11.12 A schematic of tactile sensor and rectifier circuit for detecting the direction of core movement. [Redrawn from Khodambashi (2008),[27] courtesy of Science Publications.]

difference becomes negative. Thus, the magnitude of potential difference indicates the amount of motion, and its sign indicates the direction of motion.

Figure 11.13 shows the different parts of the manufactured sensor. The core consists of three different pieces, two of which are made of bronze and one which is made of iron. The bronze pieces are attached to the ends of the iron core and act as holders. The body of the sensor is also made from bronze. Bronze is selected because its permeability is low and does not affect the flux that is passing through the coils. To decrease the friction between the core and the body, two linear bearings have been placed at the ends of the central hole.

Lesion Model Preparation To test the sensor's ability to distinguish between different skin lesions, artificial lesions have been prepared which are similar in behavior to the actual lesions. The model is shown in Fig. 11.14. This model is made up of a rigid cylindrical container which is blocked from one side, and an elastic membrane which is put on the other side. The pressure of the air inside the container is controlled through the pipe connected to it. This container is put in a larger rectangular container, as shown in Fig. 11.14, and its surrounding is filled with paraffin gel which has mechanical properties similar to the skin. The sensor is moved with the aid of a robot from point A to point B in Fig. 11.14a in such a way that its tip remains in contact with the model surface.

The membrane is formed such that when the inside pressure of the container is equal to the pressure of surrounding air, it will stand as in position 1 in Fig. 11.14b. In this case, if the sensor tip is moved on it, it will deform to position 2. This behavior is similar to the behavior

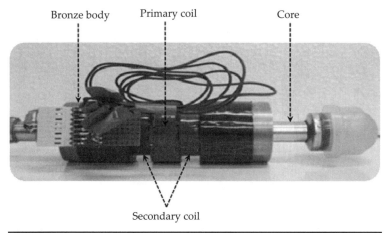

FIGURE 11.13 Different parts of the manufactured sensor. [Redrawn from Khodambashi (2008),[27] courtesy of Science Publications.]

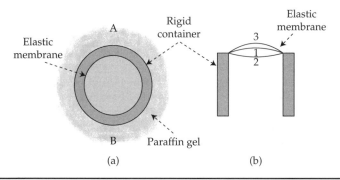

FIGURE 11.14 Schematic of the skin lesion model including (a) top view and (b) side view. [Redrawn from Khodambashi (2008),[27] courtesy of Science Publications.]

of anetoderma. If the pressure is increased above the surrounding pressure, the membrane will deform to position 3. In this case, it will deform to position 2 if the sensor's tip is moved on it.

Sensor Calibration For calibrating the sensor, the output voltage of the sensor has been obtained as a function of core displacement. A micrometer has been used to move the core for 15 mm with steps of 0.02 mm. The sensor has been fixed using a hook. The core has been attached to the movable jaw of the micrometer and moves with it. The output voltage is recorded using the DAQ system.

In Fig. 11.15, the output voltage of the sensor has been plotted as a function of core displacement. For a position sensor, linearity

Application of Tactile Sensing in Robotic Surgery

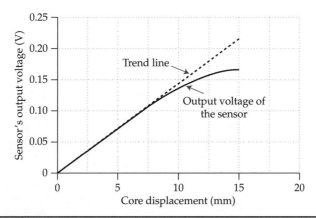

FIGURE 11.15 The output voltage of the sensor as a function of core displacement. [Redrawn from Khodambashi (2008),[27] courtesy of Science Publications.]

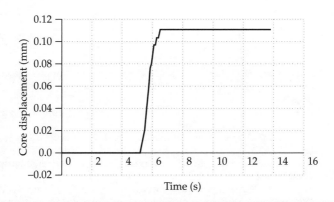

FIGURE 11.16 Core displacement as a function of time while the device crosses over a 100 microns thick paper. [Redrawn from Khodambashi (2008),[27] courtesy of Science Publications.]

is defined as the percentage of the full motion range for which this curve is a straight line. For this sensor, the curve could be considered as linear until the core moves 8 mm from the center. The full motion range is 15 mm and, thus, the linearity is 53.3%. In the linear range, the slope of the curve is 0.0145 V/mm and displacement of the core is obtained by Eq. 11.2:

$$d = \frac{V}{0.0145} \tag{11.2}$$

In the above equation, d is the displacement of the core and V is the output voltage of the sensor.

Results In order to test the sensitivity of the sensor in detecting fine surface features, it was moved over a piece of paper with thickness of 100 microns. The sensor's output voltage is converted to displacement values using Eq. 11.2, and is shown in Fig. 11.16. As can be seen, the sensor's output shows a sudden increase when it crosses the edge of paper. The thickness of the paper measured by the sensor is 110 microns, which has an error of 10%.

The sensor was set at zero output. When the sensor is put on the position A of the lesion model shown in Fig. 11.14, the output of the sensor is 0.12 V. Then the sensor was moved to position B, and at this location the output is also 0.12 V. However, when the sensor is exactly on the lesion, the output is 0.08 V; therefore, we see a drop 0.04 V when passing on lesion. This is because of the mechanical property of the simulated lesion (anetoderma). In this case, the surrounding tissue simulated with paraffin gel is stiffer than the lesion.

References

1. B. Davies, "A Review of Robotics in Surgery," Proceedings of the Institution of Mechanical Engineers. Part H, *Journal of Engineering in Medicine*, vol. 214, no. 1, 2000, pp. 129–140.
2. R. H. Taylor and D. Stoianovici, "Medical Robotics in Computer-Integrated Surgery," *Robotics and Automation*, IEEE Transactions, vol. 19, no. 5, 2003, pp. 765–781.
3. R. D. Howe and Y. Matsuoka, "Robotics for Surgery," *Annual Review of Biomedical Engineering*, vol. 1, 1999, pp. 211–240.
4. R. H. Taylor and S. D. Stulberg, "Medical Robotics Working Group Section Report," in *NSF Workshop on Medical Robotics and Computer-Assisted Medical Interventions (RCAMI)*, 1996, Bristol, England: Shadyside Hospital, Pittsburgh, Pa.
5. H. Kang, "Robotic Assisted Suturing in Minimally Invasive Surgery," Ph.D. dissertation, Rensselaer Polytechnic Institute, United States, 2002. (Dissertations & Theses: A&I database, publication No. AAT 3057670.)
6. http://biomed.brown.edu/Courses/BI108/BI108_2005_Groups/04/index.html (Materials were extracted from pages: orthopedics, neurology, urology, and cardiology.)
7. C. Coo, C. MacKenzie, and S. Payandeh, "Task and Motion Analyses in Endoscopic Surgery," *ASME IMECE Conference Proceedings: 5th Annual Symposium on Haptic Interfaces for Virtual Environment and Teleoperator Systems*, Atlanta, GA, 1996, pp. 583–590.
8. A. J. Madhani, G. Niemeyer, and J. K. Salisbury, "The Black Falcon: A Teleoperated Surgical Instrument for Minimally Invasive Surgery," *Proceedings of the 1998 IEEE/RSJ International Conference on Intelligent Robots and Systems*, vol. 2, 1998, pp. 936–944.
9. http://biomed.brown.edu/Courses/BI108/BI108_2005_Groups/04/history.html
10. http://www.intuitivesurgical.com/products/davinci_surgicalsystem/index.aspx
11. http://library.thinkquest.org/03oct/00760/Zeus System.htm
12. http://www.robodoc.com/ (Materials were extracted from old robodoc.com site: robodoc, orhtodoc, and neuromate.)

13. T. Ortmaier, H. Weiss, and V. Falk, "Design Requirements for a New Robot for Minimally Invasive Surgery," *The Industrial Robot*, vol. 31, no. 6, 2004, p. 493.
14. Ch. R. Wagner, "Force Feedback in Surgery: Physical Constraints and Haptic Information," Ph.D. dissertation, Harvard University, United States, 2006. (Dissertations & Theses: A&I database, publication No. AAT 3217920.)
15. T. A. Chase, "Design and Implementation of Capacitive Tactile Array Sensors for Dexterous Robot Fingers," Ph.D. dissertation, North Carolina State University, United States, 1997. (Dissertations & Theses: A&I database, publication No. AAT 9736816.)
16. D. Ruspini, K. Kolarov, and O. Khatib, "Haptic Interaction in Virtual Environments," *IEEEIRSJ International Conference on Intelligent Robots and Systems*, Grenoble, France, 1997.
17. M. Agus, A. Giachetti, E. Gobbetti, G. Zanetti, N.W. John, and R. J. Stone "Mastoidectomy Simulation with Combined Visual and Haptic Feedback," In *Medicine Meets Virtual Reality*, 2002, pp. 17–23.
18. http://en.wikipedia.org/wiki/telemedicine
19. Rosen, Theodore, Marilyn B. Lanning, and Marcia J. Hill, *The Nurse's Atlas of Dermatology*, Boston: Little Brown & Co., 1983.
20. G. M. White and N. H. Cox, "Structural Disorders of the Skin and Disease of Subcutaneous Tissues; Structural Disorders of the Dermis, Collagen, and Elastic Tissue," *Diseases of the Skin*, 2d ed, Elsevier Publishers, 2006.
21. M. Tanakaa, J. Hiraizumia, J. L. Leveque, and S. Chonan, "Haptic Sensor for Monitoring Skin Conditions," *International Journal of Applied Electromagnetics and Mechanics*, vol. 14, 2001–2002, pp. 397–404.
22. F. S. Mair, A. Haycox, C. May, and T. Williams, "A Review of Telemedicine Cost-Effectiveness Studies," *Journal of Telemedicine and Telecare*, Issues in Remote Diagnosis and Management of Cutaneous Disease," *Current Problems in Dermatology*, vol. 14, 2002, pp. 38–40.
23. C. M. Philips, D. Balch, S. Schanz, and A. Branigan, "Teledermatology: Issues in Remote Diagnosis and Management of Cutaneous Disease," *Current Problems in Dermatology*, vol. 14, 2002, pp. 38–40.
24. J. D. Whited, "Teledermatology Research Review," *International Journal of Dermatology*, vol. 45, no. 3, 2006, pp. 220–229.
25. N. Eminovi, N. F. de Keizer, P. J. E. Bindels, and A. Hasman, "Maturity of Teledermatology Evaluation Research: A Systematic Literature Review," *British Journal of Dermatology*, vol. 156, no. 3, 2007, pp. 412–419.
26. R. B. Mallett, "Teledermatology in Practice," *Clinical & Experimental Dermatology*, vol. 28, no. 4, 2003, pp. 356–359.
27. R. Khodambashi, S. Najarian, A. Tavakoli Golpaygani, A. Keshtgar, and Sh. Torabi, "A Tactile Sensor for Detection of Skin Surface Morphology and Its Application in Telemedicine Systems," *American Journal of Applied Sciences*, vol. 5, no. 6, 2008, pp. 633–638.
28. N. H. Cox, "A Literally Blinded Trial of Palpation in Dermatologic Diagnosis," *Journal of American Academy of Dermatology*, vol. 6, 2007, pp. 949–951.
29. L. W. Chao, T. F. Cestari, and L. Bakos, "Evaluation of an Internet-Based Teledermatology System," *Journal of Telemedicine and Telecare*, vol. 9, 2003, pp. 9–12.
30. C. Massone and H. P. Soyer, "Feasibility and Diagnostic Agreement in Teledermatopathology Using a Virtual Slide System," *Human Pathology*, vol. 38, 2007, pp. 546–554.
31. H. S. Pak, "Teledermatology and Teledermatopathology," *Seminars in Cutaneous Medicine and Surgery*, vol. 21, 2002, pp. 179–189.

CHAPTER 12

Haptics Application in Surgical Simulation

12.1 Virtual Reality (VR) and Virtual Environments (VEs)

To provide a unified workspace, we can make use of virtual environments (VEs). It allows almost complete functionality without requiring that all functions occupy the same physical space. It is defined as "... interactive, virtual image displays enhanced by special processing and by nonvisual display modalities, such as auditory and haptic, to convince users that they are immersed in a synthetic space."[1] In other words, a virtual world can be described as an application that lets users navigate and interact in real time with computer-generated 3D environments. It consists of three major elements: interaction, 3D graphics, and immersion.[2,3]

Virtual reality is both a hardware system and an emerging technology, allowing a user to interact with a computer-simulated environment. It has changed the way in which individuals interact with computers. Another description for VR is "... a fully three-dimensional computer-generated 'world' in which a person can move about and interact as if he/she actually were in an imaginary place."[3] To accomplish this, we should immerse the person's senses using VE display devices such as: head-mounted displays (HMDs); spatially immersive displays (SIDs); and desktop stereo displays, such as a responsive workbench.[1] HMDs or SIDs are used when immersion within a space is a performance requirement and workbench displays are used when viewing a single object or set of objects from a third-person point of view.[4]

The main feature that characterizes VR as being different from interactive computer graphics or multimedia is the user immersion in a synthetic environment. VR applications may commonly differ fundamentally from those associated with graphics and multimedia systems.[2] According to Bricken,[5] the essence of VR is the inclusive relationship between the participant and the virtual environments (VEs), where direct experience of the immersive environment constitutes communication. It is believed that VR constitutes the leading edge of a general evolution of present communication interfaces involving television, computer, and telephone.[6] The full immersion of the human sensorimotor channels into a vivid and global communication experience is the main characteristic of this evolution.[7] In the field of biomedical engineering, telemedicine is principally involved with transmitting medical information. Here, VR has the potential to enhance the telemedicine experience. The two principal ways in which VR can be applied are through interface and environment. As interface, they enable us to have a more intuitive manner of interacting with information. As environment, they enhance the feeling of presence during the interaction.[2]

Applications of Virtual Reality

The diverse applications of this exciting field can be categorized as follows:[8]

1. Medicine and healthcare: This includes surgical procedures such as: remote surgery or telepresence, augmented or enhanced surgery, and planning and simulation procedures before surgery. It also includes: motion analysis, rehabilitation and physical therapy, and testing of new drugs.
2. Science: This includes interactive simulations of scientific phenomena and studying molecular structure and interactions without actually mixing chemicals.
3. Education and training: This includes flight simulators for pilots, virtual classrooms, and battlefield simulations.
4. Entertainment: This includes video games, movies, and television shows.

Other fields of application of VR include: robotics, military, sports and fitness, tools for the handicapped, art, architecture, and business.[8]

Advantages and Limitations

In the areas of improved services and savings in material resources, remote virtual environments and related technologies have added value to healthcare. Improved surgical results are obtained by virtual reality. Some of the examples are: the use of laparoscopic simulators for training;[9] the development of applications that simulate human response to medication; simulator systems helping to train anesthesiologists;[10] and the development of imaging tools that guide surgical tools through brain tissue to the site of a tumor.[11] Another advantage is that we can save precious resources, such as cadavers and animals, through the use of simulators. We can also use simulators to train medical personnel. In this way, one can decrease the demand for nonrenewable resources. In other words, without reducing the supply of nonrenewable resources, the trainee can practice many times using a realistic, virtual environment.

Despite their promising potential, current virtual environment applications in healthcare are somewhat limited. Some of these technical problems are the cost, the lack of reference standards, and the noninteroperability of systems.[2] In most medical simulations, haptic feedback is coupled with graphical simulation. This is a well-known bottle neck in the area of medical robots and computer-assisted surgery and training. To obtain realistic force feedback, haptic systems need high simulation rates of about 1000 Hz.[2,12] However, the update rates of the physical objects being simulated are normally in the order of 20 Hz to 30 Hz.[12] An oscillatory behavior in the haptic device is sometimes observed due to this difference in frequency. As a result, this can inflict harm on the operator since it becomes unstable.

12.2 Haptics-Based Surgical Simulation

Haptics is the science of applying tactile sensation to computer applications in order to enable users to receive feedback, especially force feedback, in the form of sensations that they actually feel. Haptics technology is mostly used to train hand-eye coordination in tasks such as minimally invasive surgery and spaceship maneuvers.

In surgical simulations, haptic modeling includes tasks that require the sense of touch and need to provide the user with the force feedback. Palpating, pulling, cutting, tearing, puncturing, and stapling are the imperative procedures needing the sense of touch.[8]

Haptic simulation of medical procedures is an active area of research in biomedical engineering. Similar to flight simulators for pilots, medical simulators allow students to learn and practice delicate procedures that are performed on internal organs of the body. Additionally, all of these happen in a risk-free virtual environment with both visual and force feedback.[2] So far, procedures such as suturing,[13] bone dissection,[14,15] laparoscopy,[16,17] and needle insertion[2,6] have been developed.

The real-time simulation of soft tissue deformation is one of the main challenges in developing medical training simulators.[18] At present, most surgical simulations assume that soft tissue is comprised of elastic material with either uniform or varying stiffness.[2,18] If we accept this assumption, then we will be ignoring some anatomical structures such as blood vessels, nerves, or cysts. However, we know that these structures have a considerable impact on how the surrounding tissue deforms. It is possible to include them in surgical simulations, however. This is achieved by simulating fluid-filled structures enclosed in elastic media.[2] By so doing, the realism of the simulation is increased. Another important factor in medical simulations is real-time force feedback. This type of feedback can provide a great source of information for the trainees. Additionally, it allows them to detect transitions between organs and cavities. Also, this force feedback can help the trainees to identify the nature of the organs and tissue properties.[2]

Medical Training Simulation

Surgeons are trained through apprenticeship. Here, one of the main goals of medical training simulation is to provide a virtual-reality environment. Within this environment, surgeons can practice medical procedures such as using needles, scalpels, or laparoscopic tools. In fact, surgical simulation aims to develop an alternate training medium for surgery in the form of a virtual environments based surgical training simulator. The purpose of these training simulator systems is to allow students of surgery to interact with similar tools and observe their effect on the tissue in an interactive 3D simulation environment.

Generally, medical simulation systems have three basic components. They have the haptic interface (physical model), real-time simulation (computer), and graphic display (interface). These components are discussed in the following section:[2]

1. **Haptic Interface**: The haptic device is the hardware with which the user physically interacts. These devices are instrumented in such a way that the position and orientation of the tool can be calculated. Also, in order to display force feedback, some, or all, of the joints are actuated. This happens when the user interacts with the virtual environment. Due to the sensitivity of human touch, the force feedback must be updated at a high frequency. For free arm movement, a value between 300 to 500 Hz is required.

2. **Real-Time Simulation**: As the user moves the input device, the real-time simulation is responsible for computing the interaction physics between virtual tools and objects. The real-time simulation unit for user observation performs computational algorithms for collision detection, deformation, and dynamics.

3. **Graphic Display**: In certain cases, we also need visual feedback. A monochrome 2D display is used in most applications. Using stereo glasses (stereoscopic viewing) to provide 3D visual feedback is another way for display. This approach is used in the ZEUS[19] and daVinci surgical robotics suite.

By working together, these components allow the user to do the following tasks: to interact with a virtual object, to feel the simulated physical forces through the haptic interface, and to observe the motion and deformation of the object on the graphic display. In this process, graphics and haptics are updated at different rates. Haptic interfaces require much higher update rates than graphic display;[20] however, we need to synchronize them. Therefore, the touch and visual signals are combined to yield a convincing feeling that the user is actually deforming the object on screen.[2]

The flow of information in a virtual environment forms a closed loop, as illustrated in Fig. 12.1. Based on the user's input, the model deforms. The reason for this deformation is the existence of a deformation force. When this force is exerted to the user by the haptic interface, the loop is closed.

Deformable Models for Tissue Simulation

Deformable models are the most important part of a surgical simulator. There are various methods available for computing deformations: finite element method (FEM), boundary element method (BEM), method of finite spheres, mass-spring model or mass-spring-damper, hybrid models, and long element method (LEM).[8,21] FEM and BEM are elasticity-based methods. They both use continuum mechanics to describe the deformation of an elastic body. Using either volumetric (FEM) or surface (BEM) elements, this elastic body is being discretized. These methods are well suited for accurate simulation of tissue. This is because experimental elasticity parameters can be readily implemented in the model.[2] One of the disadvantages of these methods is that they are very computationally expensive. Additionally, for real-time applications, they need pre-computation and numerical optimization.

Mass-spring model is very efficient. This is mainly due to the simplicity of the motion equations. The two deficiencies of this method are that the deformations are very sensitive to the spring

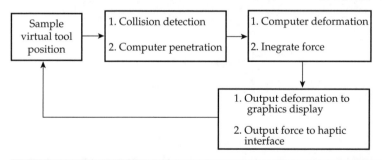

FIGURE 12.1 Data flow in a typical haptic virtual environment. [From Yang (2005),[2] used by permission.]

parameters, and that achieving tissue-like behavior is difficult.[2] The mass-spring model consists of a square grid of particles, each with a unit mass and spaced at unit distances. The masses are held in a grid by three different kinds of simulated springs. These are bend springs or flexion springs, shear springs, and structural springs.[8]

Haptic Simulation

A simplified description of virtual reality simulation for haptic output is given in Fig. 12.2. In order to feel the feedback force, a mechanical device, often called *haptic display* or *haptic interface*, is needed. This device interacts physically with the human operator who is mediating with the reaction forces of the virtual environment.

In this section, we take a look at the systems which simulate the manipulation of virtual objects through instruments. Here, the haptic interface represents a virtual instrument. This interface should have a similar handle as the device it intends to simulate. Additionally, it should accept the commands given by the user, imposing on the

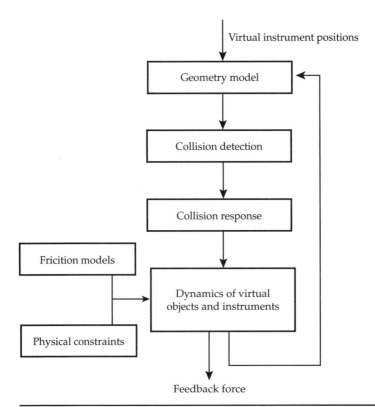

FIGURE 12.2 Virtual environment simulation. [From Yang (2005),[2] used by permission.]

human at the same time the forces are due to the interaction of the virtual instrument with the virtual environment. Based on this discussion, a dual task is conducted by the haptic interface. First, by providing motion and interaction commands, it is an input to the simulation system; second, it provides the user with kinesthetic information about the virtual environment.[2] This reflects the inherent bidirectionality of a kinesthetic interaction. This system requires an exchange of mechanical power flow. One approach is through the product of instantaneous contact force and velocity. Therefore, we can have two substantially different types of haptic interfaces. In the first type, called *impedance control* or *force feedback control*,[22] we measure the positions or velocities imposed by the user on the haptic device. In addition to this, the reaction forces are returned. In the second type, called *admittance control* or *position-feedback control*,[23] we sense the forces applied by the user. Using the haptic device, positions are fed back. While the first control scheme has been used in most systems, the second one is rarely used.

Fluid Simulation

It is quite common in medical procedures to interact with fluids as well as solid materials. Fluid physics must be computed using fluid mechanics analysis in order to model fluid-filled structures. Blood vessels or glands belong to this category. Using either static or dynamic analysis, we can solve fluid mechanics. Time-varying quantities are not included in static fluid analysis. Unlike fluid statics, fluid motion is considered in fluid dynamics.[2]

In medical simulations, both static and dynamic fluid analyses have been used. Using fluid dynamics, any fluid simulation that requires flow, such as bleeding[24,25] or removal of water and bone paste,[26] can be modeled. The main disadvantage of using fluid dynamics is that it is computationally expensive. Due to this drawback, fluid dynamics is solved only for graphical presentation. These types of presentations require a much slower update rate than for haptics. If we intend to obtain the high update rates necessary for haptics, it is possible to use fluid statics to model fluid-filled structures. These structures could be blood vessels or other organs.[27]

Surgical Simulators Based on Haptics

Traditionally, medical students practice on animals, cadavers, mannequins, and patients by observing an experienced surgeon. However, animals do not have the same anatomy as humans, and practicing on them can be very expensive. Ethical issues can also be raised when using animals. Additionally, we know that even cadavers cannot provide the correct physiological response, and practicing on mannequins is unrealistic and inflexible. Therefore, while the student gains competence, there is a risk to patient safety. To tackle these issues, haptics and VR-based surgical simulators may be quite useful.[2,21] Their role is to provide a safe and viable alternative. Realistic human anatomy can be incorporated into virtual patient models. At the same time, we can simulate both normal and pathological physiology. In addition, simulators can provide a structured learning environment with controlled difficulty levels. It is believed that for realistic training in surgery, haptic feedback is quite necessary. So far, relatively little work has been done on haptic perception. However, various methods are available for haptic simulation in surgical applications. In general, they can be divided into four groups according to their force feedback complexity. These groups are needle-based,[2,28,29] open surgery, telesurgery,[20] and minimally invasive surgery.

Needle-Based Procedure

As the name suggests, needle-based procedures use needles, catheters, guide wires, and other small bore instruments.[2]

The Immersion Medical CathSim Vascular Access Simulator[30] was developed to train nursing students in the proper technique for starting an intravenous line. To simulate the needle and catheter, this system used the CathSim AccuTouch device[30] as the haptic interface device. It consists of an enclosed needle carrier with various sensors located on it. It provides pitch, yaw, and depth. Detection of the depth to which cannula or guide wire has been inserted is done by encoders located on the carrier. A variable amount of braking force can be applied, depending on whether the carrier is pushed or pulled. Three degrees of freedom (DOF) orientation data can be reported by the device. It also has one DOF haptic feedback along the direction of needle insertion. The same hardware configuration was used by Liu et al.[31] to develop needle-based trauma procedures. These procedures have been successfully employed in pericardiocentesis and diagnostic peritoneal lavage operations.

Needle-based simulators have limited haptic realism. They can have an impact for training widely performed procedures at low cost, in situations where opportunities for practice are limited, and where current methods using animal models are not optimal.[2]

Despite recent advances in surgical simulators for MIS, limitations still exist. Open surgery is more difficult than MIS to simulate. This is true because the visual field, the range of haptic feedback, and the freedom of motion are considerably larger compared to minimally invasive procedures.[2] Having mentioned the facts, open surgery still remains the Holy Grail of surgical simulation. This means that before open surgery can be simulated realistically, we have to make considerable advances in haptics, real-time deformation, organ and tissue modeling, and visual rendering.[2]

Commercially available laparoscopic trainers (LapSIM—the laparoscopic training tool and Mentice—medical simulators) can teach basic skills. These include camera navigation, grasping, suturing and knot tying, and cauterization. To improve performance in the operating room, laparoscopy simulators have been clinically validated. Like other similar applications, and despite recent advances, limitations still exist. This is because real-time tissue and organ deformations are generally limited to either specific organs or simple structures such as arteries, ducts, and other tubular structures.[2]

References

1. S. R. Ellis, "What Are Virtual Environments?," IEEE Computer Society Press, vol. 14, no. 1, 1994, pp. 17–22.
2. H. Yang, "Modeling of Needle Insertion Forces for Haptics-Based Surgical Simulation," M.A.Sc. dissertation, Carleton University, Canada, 2005. (Dissertations & Theses: A&I database, Publication No. AAT MR08392.)
3. R. M. Satava, "Virtual Reality Surgical Simulator: The First Steps," *Surg Endosc.*, vol. 7, no. 3, 1993, pp. 203–205.
4. K. M. Stanney, *Handbook of Virtual Environments: Design, Implementation, and Applications*, Lawrence Erlbaum Associates, 2002.
5. S. Cotin and H. Delingette, "Real-Time Surgery Simulation with Haptic Feedback Using Finite Elements," *IEEE International Conference on Robotics and Automation*, 1998.
6. S. P. DiMaio and S. E. Salcudean, "Simulated Interactive Needle Insertion," *10th Symposium on Haptic Interfaces for Virtual Environment and Teleoperator Systems*, 2002.
7. I. Brouwer, K. E. MacLean, A. J. Hodgson, A. G. Nagy, K. A. Qayumi, and L. M. Rucker, "Price Quality Trade-offs in Haptic Interfaces for Simulation of Laparoscopic Surgery," in Poster Presented at Medicine Meets Virtual Reality, 2003.
8. G. Gopalakrishnan, "Stapsim: Virtual Reality-Based Stapling Simulation for Laparoscopic Herniorrhaphy," M.A.Sc. dissertation, The University of Texas at Arlington, United States, 2003. (Dissertations & Theses: Full Text database, Publication No. AAT 1414742.)
9. M. S. Wilson, A. Middlebrook, C. Sutton, R. Stone, and R. F. McCloy, "MIST VR: A Virtual Reality Trainer for Laparoscopic Surgery Assesses Performance," *Annals of the Royal College of Surgeon of England*, vol. 79, no. 6, 1997, pp. 403–404.
10. D. E. Burt, "Virtual Reality in Anesthesia," *British Journal of Anesthesia*, vol. 75, 1995, pp. 472–480.
11. C. Giorgi, M. Luzzara, D. S. Casolino, and E. Ongania, "A Computer Controlled Stereotactic Arm: Virtual Reality in Neurosurgical Procedures," *Acta Neurochir Suppl (Wien)*, vol. 58, 1993, pp. 75–76.
12. H. Z. Tan, M. A. Srinivasan, B. Eberman, and B. Cheng, "Human Factors for the Design of Force-Reflecting Haptic Interface," *Proc. Winter Annual Meeting of the American Society of Mechanical Engineers: Dynamic Systems and Control*, vol. 55–1, 1994, pp. 353–359.
13. A. J. Lindblad, "Increasing the Functionality of Finite Element Based Surgical Suturing Simulators," Ph.D. dissertation, University of Washington, United States, 2006. (Dissertations & Theses: A&I database, Publication No. AAT 3241925.)
14. D. Morris, "Haptics and Physical Simulation for Virtual Bone Surgery," Ph.D. dissertation, Stanford University, United States, 2006. (Dissertations & Theses: A&I database, Publication No. AAT 3235299.)

15. M. Agus, A. Giachetti, E. Gobbetti, G. Zanetti, and A. Zorcol, "Real-Time Haptic and Visual Simulation of Bone Dissection," *IEEE Virtual Reality*, 2002, pp. 209–216.
16. C. Basdogan, C. H. Ho, and M. A. Srinivasan, "Virtual Environments for Medical Training: Graphical and Haptic Simulation of Laparoscopic Common Bile Duct Exploration," *IEEE/ASME Transactions on Mechatronics*, vol. 6–3, 2001, pp. 269–285.
17. V. Chandramouli, "A Hybrid Collision Detection Method for Virtual Laparoscopic Surgery," M.S. dissertation, The University of Texas at Arlington, United States, 2003. (Dissertations & Theses: A&I database, Publication No. AAT 1418158.)
18. H. Shi, "Finite Element Modeling of Soft Tissue Deformation," Ph.D. dissertation, University of Louisville, United States, 2007. (Dissertations & Theses: A&I database, Publication No. AAT 3293566.)
19. http://www.computermotion.com/productsandsolutionslproductslzeuslindex.cfin.
20. M. C. Cavusoglu, "Telesurgery and Surgical Simulation: Design, Modeling, and Evaluation of Haptic Interfaces to Real and Virtual Surgical Environments," Ph.D. dissertation, University of California, Berkeley, United States, 2000. (Dissertations & Theses: A&I database, Publication No. AAT 3001778.)
21. X. Wang, "Deformable Haptic Models for Surgical Simulation," Ph.D. dissertation, The University of Texas at Arlington, United States, 2005. (Dissertations & Theses: A&I database, Publication No. AAT 3181891.)
22. T. H. Massie and J. K. Salisbury, "The Phantom Haptic Interface: A Device for Probing Virtual Objects," *American Society of Mechanical Engineers, Dynamic Systems and Control Division*, vol. 55–1, 1994, pp. 295–301.
23. H. Yokoi, "Development of the Virtual Shape Manipulating System," *Proceedings of the Fourth International Conference on Artificial Reality and Tele-Existence*, Tokyo, 1994, pp. 43–48.
24. L. Raghupathi, "Simulation of Bleeding and Other Visual Effects for Virtual Laparoscopic Surgery," M.S.E.E. dissertation, The University of Texas at Arlington, United States, 2002. (Dissertations & Theses: A&I database, Publication No. AAT 1412849.)
25. H. Yokoi, "Development of the Virtual Shape Manipulating System," *Proceedings of the Fourth International Conference on Artificial Reality and Tele-Existence*, Tokyo, 1994, pp. 43–48.
26. M. Agus, A. Giachetti, E. Gobbetti, G. Zanetti, N. W. John, and R. J. Stone, "Mastoidectomy Simulation with Combined Visual and Haptic Feedback," *Medicine Meets Virtual Reality*, 2002, pp. 17–23.
27. S. De and M. A. Srinivasan, "Thin Walled Models for Haptic and Graphical Rendering of Soft Tissues in Surgical Simulations," *Medicine Meets Virtual Reality*, 1999, pp. 94–99.
28. S. P. DiMaio and S. E. Salcudean, "Simulated Interactive Needle Insertion," *Proc. 10th Symposium on Haptic Interfaces for Virtual Environment and Teleoperator Systems*, 2002, pp. 344–351.
29. S. P. DiMaio, "Modelling, Simulation and Planning of Needle Motion in Soft Tissues," Ph.D. dissertation, The University of British Columbia, Canada, 2003.
30. M. Ursino, P. D. J. L. Tasto, B. H. Nguyen, R. Cunningham, and G. L. Merril, "CathSimTM: An Intravascular Catheterization Simulator on a PC," *Medicine Meets Virtual Reality*, MMVR 7, 1999, pp. 360–366.
31. A. Liu, C. Kufinann, and D. Tanaka, *An Architecture for Simulating Needle-Based Surgical Procedures*, MICCAI 2001, pp. 1137–1144.

Abbreviations

STS	static tactile sensing
DTS	dynamic tactile sensing
X	variable (input)
x	variable (input)
S	variable (output)
f	transfer function or local force tensor
FS	input full scale
T_r	response time
T_d	decay time
T_{on}	onset time
T_{off}	offset time
C	capacitance or pyroelectric coefficient
ε_0	permittivity
ε_r	dielectric constant
A	area
d	distance or diameter
MUX	multiplexor
SP	static plate
MP	moving plate
mmf	magnetomotive force
Φ	magnetic flux
B	magnetic flux density
L	length
μ	permeability
μ_R	relative permeability
LVDT	linear variable differential transformer
V	volt (output) or volume
LED	light emitting diode
PIN	photodiode
V_H	Hall voltage
PET	polyethylene terephthalate
R_0	resistance in unstrained state
P	resistivity
l	length or tissue loading
ε	strain or dielectric constant
v	Poisson's ratio
F	force
GF	gauge factor
E	elastic modulus (Young's modulus)
Δ	deflection
B	width
M	bending moment
σ	stress

Abbreviations

h	height or depth
I	moment of inertia
λ_1	longitudinal piezoresistive coefficient
c_1	constant
c_2	constant
g	directional piezoelectric constant
PVDF	polyvinylidene fluoride
t	thickness or time
T	absolute temperature
MIS	minimally invasive surgery
MEMS	micro-electro-mechanical systems
TMAH	tetramethylammonium hydroxide
DI	deionization
MAS	minimal access surgery
IC	integrated circuit
k	spring constant
η	viscosity constant
J_1	creep compliance
G	shear relaxation function
P	pressure
BSE	breast self examination
CBE	clinical breast examination
u	displacement
FEM	finite element method
E_T	tissue modulus
E_L	lesion or tumor modulus
CTI	composite tactile image
p	pressure
z	depth
Er	stiffness ratio
s	tumor shape
ANN	artificial neural network
FSR	force sensing resistor
PTF	polymer thick film
TPDT	two-point discrimination threshold
AFM	atomic force microscopy
OPD	out-of-plane deflection
TIT	tissue including a tumor
TIHA	tissue including a healthy artery
TISA	tissue including a stenotic artery
PDMS	polydimethylsiloxane
CIS	computer-integrated surgery
CT	computer tomography
MRI	magnetic resonance imaging
THA	total hip arthroplasty
TKA	total knee arthroplasty
CABG	coronary artery bypass grafting
OPCAB	Off-pump Coronary Artery Bypass Grafting
TURP	transurethral resection of the prostate
MIRS	minimally invasive robotic surgery
SAF	store-and-forward
RT	real-time
HMD	head-mounted display
SID	spatially immersive display
VE	virtual environment
VR	virtual reality
BEM	boundary element method
LEM	long element method

Index

A

accommodation, 2
accuracy, 41–42
Accuray Incorporated, 205
achieved reliability, 43
action potentials, 27, 28f, 29f
action potentials-per-second (APs/Sec), 28, 29f
active sensing, 35–36
adipose tissue, 130
admittance control, 227
AESOP (A voice-controlled endoscope positioning robot), 205, 206–207
afferent neurons, 30–31, 31f
AFM. *See* atomic force microscopy (AFM)
air pressure, 8
amacrine cells, 6
ambient temperature, 80
amplitude, 8
angioplasty, 201
angioscopy, 106
animals, sense of touch in, 19
annulospiral endings, 25, 26f
anosmia, 15
ANSYS software, 152
appendectomy, 106
arteries, detection of, 184–193
arthroscopy, 106
artificial neural networks (ANNs), for estimation of tumor characteristics, 149–152
artificial sensors
   accuracy of, 41–42
   calibration of, 40
   classification of, 44–46
   hysteresis, 40–41
   linearity of, 40
   noise, 42
   reliability of, 43
   repeatability and, 43
   resolution of, 39
   response time of, 43–44
   sensitivity of, 39–40
   span or dynamic range of, 42
   specifications for, 44
   terminology of, 38–44
   transfer function, 39
artificial skin, 193–195, 194f
artificial tactile sensing, 38
atomic force microscopy (AFM), 171
audible frequencies, 8
audio systems, xi
audition, 6–12
auditory afferents, activation of, 10–11
auditory canal, 8–9
auditory system, 7
axons, 27

B

base force/torque sensor, 204, 204f

basilar membrane, 10–11, 13f
BEM. *See* boundary element method (BEM)
bimorph configuration, 89, 89f
binary pressure sensors, 66–68, 67f
biological tissue
   contact force between endoscopic grasper and, 115–121
   schematic representation of, 187f
   tactile distinction of an artery and a tumor in, by finite element method, 184–193
biological tissue properties
   flexible membrane tactile sensor for, 173–180
   hardness, 180–182
   introduction to, 171–173
   modulus of elasticity, 182–193
   stiffness, 172–173, 183
biomedical applications. *See also* surgery
   endoscopic surgery, 92, 92f
   minimally invasive surgery, 92–94
   of piezoelectric sensors, 91–103
bipolar cells, 6
bitterness, 16–17
bonded gauges, 75, 76f, 77
boundary element method (BEM), 225
brain
   construction of images by, 6, 7f
   tactile information processing by, 31
brain surgery, 200–201
breast
   anatomy of, 129–130, 129f
   tactile imaging of, 128–130
breast cancer
   estimating of lesion parameters, 132–133
   imaging procedures for, 130–132, 139–140
   screening and detection, 128–132
   statistics, 130
   tactile imaging and, 131–132
breast self examination (BSE), 130–131

C

calibration, 40
calibration curve, 40, 40f
capacitors, 90
capacitive sensors, 49–52, 50f, 51f, 52f
   hybrid piezoelectric-capacitive tactile sensor, 99–103, 101f
carbon fiber sensors, 59, 61–63, 61f, 62f
cardiothoracic surgery, 201
cartilage, stiffness of, 172–173
CathSim AccuTouch device, 228
cells
   amacrine, 6
   bipolar, 6
   cone, 4–6, 7f
   ganglion, 5, 6
   horizontal, 5
   load, 78
   photoreceptor, 2, 4, 5
   receptor, 27–28, 27f
   sensory, 1, 28f
chemoreceptors, 21
cholecystectomy, 106
chondrocytes, 172
cilia, 10
ciliary muscles, 2–4, 4f
clinical breast exam (CBE), 131, 143–144
closed-chest heart bypass surgery, 199
cochlea, 9–10
coherent radiation, 171
colectomy, 106
colonoscopy, 106

Index

color vision, 5, 6
commercial surgical robots, 205–208
communication, by skin, 21
computational tactile sensing method, for tumor detection, 143–149
Computer Motion, 205
computer tomography (CT), 200
conductive elastomer sensors, 59–61, 60f, 61f
cone cells, 4–6, 7f
Constantan, 75
contact force, between endoscopic grasper and biological tissues, 115–121
contact-force estimation, 174, 174f
coordinate systems, 87, 87f
copper-nickel alloy, 75
cornea, 4f
coronary artery bypass grafting (CABG), 201
cortex, 6, 7f
crystallins, 2–3
crystals, 85, 86
Curie, Jacques, 85
Curie, Pierre, 85
Curie temperature, 85–86
cutaneous sensing, 35, 112–113
CyberKnife, 205

D

data acquisition interfaced (DAQ), 155
data processing, 109–111
da Vinci, 205, 206, 208
DC blocking filter, 90–91
decay time, 43–44, 43f
decibels, 42
deformable models, for tissue simulation, 225–226
demodulator, 57
derivative information, 45
dermis, 22–23, 22f
desktop stereo displays, 221
drugs, hearing loss and, 12
dynamic range, 42
dynamic tactile sensing (DTS), 35–36
dynamic tactile sensors, 45–46

E

eardrum, 9
ears
    function of, 6–10
    structure of, 9f, 10f
elasticity, measuring modulus of, 182–193
Elastirob, 182–184, 185f
elastography, 171
endoscopic grasper, 82–83, 83f, 84f
    contact force between biological tissues and, 115–121
    with graphical display of tactile sensing data, 94–99, 95f, 96f
    for graphical rending of localized lumps, 153–169
endoscopic surgery, 105
    piezoelectric sensors in, 92, 92f
endosurgery, 105
end-supported, center-driven technique, 90
epidermis, 22f
E-shape core variable air gap sensor, 55, 56f
evaporation, control of, 21
excretion, 21
exteroceptive sensor system, 22–24, 35
extracellular matrix, 172
extrinsic tactile sensors, 45–46
eyeball, shape of, 3–4
eyes
    adjustment of, 2–6
    features explored by, 3t
    function of, 1–2
    structure of, 2f

F

Faraday's induction law, 214
fast adapting (FA) receptors, 22
finger motion, 125, 126t
fingertips, sensory specifications for, 44t
finite element analysis, 135–136, 152
finite element method (FEM), 109, 225
    tactile distinction of soft tissues by, 184–193
finite element models, 135–136, 135f, 136f, 137f
finite spheres, 225
flavor detection, 16–17
flexible membrane tactile sensor, 173–180
flower spray endings, 25, 26f
fluidic coupling, 68, 68f
fluid simulation, 227
force feedback, 208–209
force feedback control, 227
force sensing resistor (FSR), 152
force sensors, 78–80, 79f, 80f, 208–209
fovea, 5
frequency, 8
friction, 46
friction knot, 202
full-scale displacement, 58

G

gallium-antimony (GaSb), 81
gallium arsenide (GaAs), 81
ganglion cells, 5, 6
gauge factor, 75, 81–82
gauge length, 75
germanium, 81
glandular tissue, 130
Golgi tendon organs, 26
graphical display
    of localized lumps, 153–169
    of tactile sensing data, 94–99, 95f, 96f
grasps, types of, 124–125
gustation, 16–17

H

Hall effect, 68–69, 69f, 70f, 71
Hall voltage (VH), 69
haptic display, 226–227
haptic interface, 226–227
haptic perception, 36, 36f
haptics, 36, 113
haptics applications, in surgical simulation, 221–229
haptic sensors, for monitoring skin conditions, 212
haptic simulation, 226–227
hardness detection, 180–182
head-mounted displays (HMDs), 221
hearing, 6–12
hearing loss, 12
heart bypass surgery, 199
heat regulation, by skin, 20–21
HERMES Control Center, 205, 207
horizontal cells, 5
hybrid piezoelectric-capacitive tactile sensor, 99–103, 101f
hyper-elasticity, 59
hysteresis, 40–41, 41f

I

Immersion Medical CathSim Vascular Access Simulator, 228
immunology, 21
impedance control, 227
incus, 9
inductive sensors, 52–59
    with coil and plunger-type armature, 53–55, 54f, 55f
infections, hearing loss and, 12
inherent noise, 42
inner ear, function of, 9–10
input full scale (FS), 42
integrated circuit (IC) technology, 109
Integrated Surgical Systems Incorporated, 205
interference noise, 42
Internet, 28

interstitial fluid, 172
intrinsic reliability, 43
intrinsic sensors, 45–46, 46f
Intuitive Surgical, 205
inversion algorithm, 136–140
involution, 129
Isoelastic, 75

J
joints, 26

K
Kant, Immanuel, 1
Kelvin Model, 115, 116f
keyhole surgery, 105
kinesthetic feedback, 124
kinesthetics, 35
kinesthetic sensing, 112–113
knot-tying techniques, 202, 202f

L
labeled lines, 30
laparoscopic suturing, 202–203
laparoscopic trainers, 228–229
laparoscopy, 106
laparotomy, 105
LapSim, 228–229
large deformation, 59
lead wire desensitization, 81
lead wire resistance, 81
lens, 2–4, 4f
light detection, 4
light emitters, 63
light emitting diode (LED), 63
light receivers, 63
linearity, 40, 58–59
linear variable differential transformer (LVDT), 55–59, 56f, 57f
load cells, 78
long element method (LEM), 225
loudness, coding of, 11–12, 13f
loud sounds, 12
lump localizers, 98–99, 98f, 100f, 124

M
magnetic field, 52, 68–69
magnetic field intensity, 71
magnetic resonance imaging (MRI), 200
magnetomotive force, 53
magnetoresistance, 69, 70f, 71
magnetoresistors, 69–71, 70f
malleus, 9
mammography, 131
mapping, tactile imaging, 125–130
mass-spring model, 225–226
master slave robots, 210, 210f
mechanical properties, measuring, 45–46
mechanical stimuli, 29f
  transduction of, to neural impulses, 27–30, 27f
mechanical transients, 46
mechanoreceptors, 21, 22, 22f
medical training simulation, 224–227
Meissner's corpuscles, 22, 22f, 23
Mentice, 228–229
Merkel's discs, 22, 22f, 23
metal foils, 75, 77
metal strain gauges, 73–81
micro-electro-mechanical systems (MEMS) technology, 92
micromachined active tactile sensor, 180–182, 181f
middle ear, function of, 9
minimal access surgery (MAS), 105
minimally invasive robotic surgery (MIRS), 208
minimally invasive surgery (MIS), 105–108, xii
  applications of, 106, 107f
  disadvantages, 107
  piezoelectric sensors for, 92–94, 93f
  remote palpation instruments for, 112–121
  robots for, 208–218

surgical instruments in, 106, 107f, 116f
suturing in, 201–204
tactile distinction of soft tissues in, 184–193
tactile sensing system for, 108–112
working environment in, 111–112
model-based robotic surgery, 200
modulation, 57
modulator, 57
moving plates (MP), 51
multiplexing, 50–51
multiplexor (MUX), 50–51, 52f
muscle tension, 26

N

needle-based procedures, 228–229
nerve endings, 25–26
nervous system, 1
neural impulses, 2, 4, 7f, 21
transduction of mechanical stimuli to, 27–30
neural networks, artificial, 149–152
NeuroMate, 205, 207
neurons, afferent, 30–31, 31f
neurosurgery, 199, 200–201
nickel-iron alloy, 75
nociceptors, 21
noise, 42
nominal linear range, of LVDT, 57–58
non-model-based robotic surgery, 200
null voltage, 59

O

odor qualities, 15f
odors, 12–13, 14–15
off-pump coronary artery bypass grafting (OPCAB), 201
old age, hearing loss and, 12
olfaction, 12–15
olfactory receptors, 13
olfactory system, 13–15, 14f
open surgery, 105, 106f, 228
ophthalmology, 199
optical sensors, 63–65, 63f, 64f
opto-mechanical touch sensors, 63
organ of corti, 10
Orthodoc, 205, 207
orthopedic surgery, 199, 200
osmoreceptors, 21
ossicles, 9
outer ear, 8–9
overload protection, 80, 80f

P

Pacinian corpuscles, 22, 22–23, 22f
palpation
basic motions, 126t
introduction to, 123–124
tactile image and, 125–140
taxonomy of, 124–125
tissues and organs diagnosed by, 127f
passive sensing, 35–36
perimenopause, 129
peripheral vision, 4
phase-sensitive demodulator, 57, 58f
photodiode (PIN), 63
photons, 2
photoreceptor cells, 2, 4, 5
photoreceptors, 21
piezoelectric ceramics, 85–86, 86f, 90f
piezoelectric effect, 85
piezoelectric materials, 85
piezoelectric sensors, 71
applications of, 90
bimorph configuration, 89, 89f
biomedical applications, 91–103
directional dependence and, 86–91
disadvantages, 90

end-supported, center-driven technique, 90
frequency response, 90–91, 90f
with graphical display of tactile sensing data, 94–99, 95f, 96f
hybrid piezoelectric-capacitive tactile sensor, 99–103, 101f
materials, 85
piezoelectric ceramics, 85–86
polyvinylidene fluoride, 91
piezoelectricity, directional dependence of, 86–91
piezoresistive sensors, 59–63
pinna, 8–9
Plexiglas, 102
plunger-type armature, 53
Poisson's ratio, 183
poling, 86
polydimethylsiloxane (PDMS) pillars, 195
polyethylene terephthalate (PET), 71
polymer thick film (PTF), 152
polyvinylidene difluoride, 91
polyvinylidene fluoride (PVDF), 91, 154
position-feedback control, 227
power grasps, 125
precision grasps, 125
primary motor cortex, 31
proprioception, 24–25
proprioceptive sensor system, 24–26, 35
protection, by skin, 20
PVDF film, 102, 212, 212f
pyroelectric effect, 91

R

radio-surgery, 205
rapidly adapting (RA) receptors, 22
receptive field, 24, 25f
receptor cells, depolarization of, 27–28, 27f
reef knot, 202
registration, 200
relative measurements, 46
reliability, 43
reluctance, 52–53
remote palpation instruments, 112f, 113f, 124
design specifications for, 114–115
for MIS, 112–121
rendering algorithm, 156–164
repeatability, 43, 59
resolution, 39, 59
response time, 43–44, 43f
retina, 2, 4
Robodoc, 205, 207
robotics, 65
robotic surgery
advantages and disadvantages of, 199
applications of, 198–208
classification of, 200
commercial robots for, 205–208
force feedback in, 208–209
integrated system of, 197, 198t
model-based, 200
non-model-based, 200
suturing, 201–204
Robot Institute of America, 197
robots, 38, 193
commercial, 205–208
definitions of, 197
master-slave, 210, 210f
for MIS, 208–218
telerobotic systems, 209–212
rod cells, 4, 7f
round window, function of, 10, 11f
Ruffini cylinders, 22, 22f, 23

S

saltiness, 16–17
self-heating, 80–81
semiconductor strain gauges, 81–84
sensation, 21
senses, 1
    hearing, 6–12
    sight, 1–6
    smell, 12–15
    taste, 16–17
    touch, 19–33
sensitivity, 39–40, 59
sensory cells, 1
    components of, 28f
sensory perception, 1
sensory receptors, 21
sensory systems
    exteroceptive, 22–24
    proprioceptive, 24–26
sight, 1–6
signal conditioning, 109
silicone-rubber, 59–60, 81
skin, 19–20, 23f
    artificial, 193–195, 194f
    functions of, 20–21
    layers, 22f
skin conditions, telemonitoring, 211–218
skin hairs, 23–24, 24f
skin lesions, 211–212, 211f
skin surface morphology, tactile sensor for detection of, 212–218
slowly adapting (SA) receptors, 22
smell, 12–15
softness sensing technique, 93–94, 94f, 96–98, 97f
soft tissue. *See* biological tissue
soft tissue deformation, simulation of, 223–224
solid mechanics, 109
somatosensory cortex, 31
sound frequencies, 11
sound waves, 7, 8, 8f, 12f
sourness, 16–17

span, 42
spatially immersive displays (SIDs), 221
speech analysis, xi
speech recognition, xi
spiders, 19
spinal cord, 31
square knot, 202
Stanford Research Institute (SRI) International, 205
stapes, 9
static plates (SP), 51
static tactile sensors, 45–46, 46f
static tensile sensing (STS), 35, 35–36
stiffness, 183
    of biological tissues, 172–173
    detection, 174, 175f
strain gauge sensors, 71, 77f
    introduction to, 73
    metal strain gauges, 73–81
    semiconductor strain gauges, 81–84
stress graphs, 144–151, 146f, 147f, 148f, 151f
subcutaneous fat, 22f
support layer, 109
surface scanning, 45
surgeon
    kinesthetic feedback for, 124
    tactile feedback for, 114, 124
surgeon's knot, 202
surgery. *See also* Minimally Invasive Surgery (MIS)
    cardiothoracic, 201
    endoscopic, 92, 105
    open, 105, 106f, 228
    robotic, 197–218
    tactile sensing in, xii
    telesurgery, 210–212
surgical applications, of tactile sensing, 105–121
surgical robots. *See also* robots
    force sensors for, 208–209
    for MIS, 208–218

surgical simulation
　haptics applications in, 221–229
　medical training simulation, 224–227
　virtual reality and virtual environments, 221–223
surgical training, 199, 224–227
suturing
　laparoscopic, 202–203
　in MIS, 201–204
　tension measurement in, 203–204
sweetness, 16–17

T

TacPlay, 182–183, 184
tactile acuity, 24
tactile cells, 50–51
tactile data processing, 109–111
tactile display, 111
tactile feedback, 114, 124
tactile image information, 123–140
　introduction to palpation, 123–124
　mapping, 125–130
　palpation and, 125–140
　taxonomy of palpation, 124–125
tactile imagers, 126–127, 128f
tactile images, 144, 145, 145f
　two-dimensional, 189f, 191–192
tactile imaging, of the breast, 128–130
tactile information
　from finite element models, 135–136
　parameter estimation from, 132–133, 133f
　pathways of, 30–31, 32f
tactile maps, 127–128, 128f, 144
tactile sensation, 35, xi
tactile sensing, 35–38, 112–113
　applications of, xi–xii
　artificial, 37–38

computational, in tumor detection, 143–149
development of, xi
introduction to, 143
process of, 37f
in robotic surgery, 197–218
special features of, 31–33
in tumor detection, 143–169
tactile sensing data, graphical display of, 94–99, 95f, 96f
tactile sensing system
　design considerations in, 111–112
　for measuring stiffness of cartilage, 172, 173f
　for measuring the modulus of elasticity of soft tissues, 182–193
　for MIS, 108–112, 109f, 110f
　tactile data processing, 109–111
　tactile display, 111
tactile sensing technologies
　binary pressure sensors, 66–68
　capacitive sensors, 49–52
　carbon fiber sensors, 59, 61–63, 61f, 62f
　conductive elastomer sensors, 59–61, 60f, 61f
　fluidic coupling, 68, 68f
　inductive sensors, 52–59
　introduction to, 49
　magnetoresistors, 69–71, 70f
　optical sensors, 63–65
　surgical applications, 105–121
　thermal sensors, 65
　time of flight sensors, 65–66
tactile sensors, 38–46
　classification of, 44–46
　for detection of skin surface morphology, 212–218

flexible membrane, 173–180
micromachined active, for hardness detection, 180–182
for MIS, 108–109, 111f
specifications for, 44
terminology of artificial, 38–44
Tactile Tumor Detector, 152–153
taction. *See* touch
taste, 16–17
taste buds, 16, 17
taste nerves, 16
taste receptor cells, 16
Tekscan sensors, 195
teledermatology, 212–218
telemanipulation, 208
telemedicine, 210–211
teleoperation, 209–212
telepresence, 208
telesurgery, 210–212
tensile pressure, 74f
tetramethylammonium hydroxide (TMAH), 99
thermal sensors, 65
thoracoscopy, 106
time of flight sensors, 65–66, 66f, 67f
tissue biomechanics, 171
tissue contrast differences, 171
tissue motion, 171
tissue simulation, deformable models for, 225–226
tongue, 17f
total hip arthroplasty, 200
touch, 19–22
touch receptors, 22, 26
transfer function, 39
transurethral resection of the prostate (TURP), 201
transverse sensitivity, 80
tumor detection, 125–126
artificial neural networks for, 149–152
graphical rendering of localized lumps, 153–169
using computational tactile sensing method, 143–149
using Tactile Tumor Detector, 152–153
two-dimensional surface texture image detection, 174
two-dimensional tactile images, 189f, 191–192

U

ultrasonic waves, 66
unbonded gauges, 75, 77–78, 78f
urology, 201

V

variable air gap sensors, 53, 54f, 56f
virtual environments (VEs), 221–223
virtual reality (VR), 221–223
advantages and limitations of, 222–223
applications of, 222
viscosity, 183
vision, 1–6
vision sensors, 204, 204f
visual cortex, 6, 7f
visual information, 1–2, 6
von Mises stress graph, 187, 188, 190f, 191f, 192f

W

Wheatstone bridge, 177, 178f

Y

Young's modulus, 130, 157, 183

Z

Zeus Robotic Surgical System, 205
Zeus surgical system, 207

R
857
.T32
N25

2009